Digital Image Processing for Ophthalmology

for Ophthalmology

Detection and Modeling of Retinal Vascular Architecture

Synthesis Lectures on Biomedical Engineering

Editor
John D. Enderle, *University of Connecticut*

Lectures in Biomedical Engineering will be comprised of 75- to 150-page publications on advanced and state-of-the-art topics that span the field of biomedical engineering, from the atom and molecule to large diagnostic equipment. Each lecture covers, for that topic, the fundamental principles in a unified manner, develops underlying concepts needed for sequential material, and progresses to more advanced topics. Computer software and multimedia, when appropriate and available, are included for simulation, computation, visualization and design. The authors selected to write the lectures are leading experts on the subject who have extensive background in theory, application, and design. The series is designed to meet the demands of the 21st century technology and the rapid advancements in the all-encompassing field of biomedical engineering that includes biochemical processes, biomaterials, biomechanics, bioinstrumentation, physiological modeling, biosignal processing, bioinformatics, biocomplexity, medical and molecular imaging, rehabilitation engineering, biomimetic nano-electrokinetics, biosensors, biotechnology, clinical engineering, biomedical devices, drug discovery and delivery systems, tissue engineering, proteomics, functional genomics, and molecular and cellular engineering.

Computer-aided Detection of Architectural Distortion in Prior Mammograms of Interval Cancer
Shantanu Banik, Rangaraj M. Rangayyan, and J.E. Leo Desautels
2013

Content-based Retrieval of Medical Images: Landmarking, Indexing, and Relevance Feedback
Paulo Mazzoncini de Azevedo-Marques and Rangaraj Mandayam Rangayyan
2013

Chronobioengineering: Introduction to Biological Rhythms with Applications, Volume 1
Donald McEachron
2012

Medical Equipment Maintenance: Management and Oversight
Binseng Wang
2012

Fractal Analysis of Breast Masses in Mammograms
Thanh M. Cabral and Rangaraj M. Rangayyan
2012

Capstone Design Courses, Part II: Preparing Biomedical Engineers for the Real World
Jay R. Goldberg
2012

Ethics for Bioengineers
Monique Frize
2011

Computational Genomic Signatures
Ozkan Ufuk Nalbantoglu and Khalid Sayood
2011

Digital Image Processing for Ophthalmology: Detection of the Optic Nerve Head
Xiaolu Zhu, Rangaraj M. Rangayyan, and Anna L. Ells
2011

Modeling and Analysis of Shape with Applications in Computer-Aided Diagnosis of Breast Cancer
Denise Guliato and Rangaraj M. Rangayyan
2011

Analysis of Oriented Texture with Applications to the Detection of Architectural Distortion in Mammograms
Fábio J. Ayres, Rangaraj M. Rangayyan, and J. E. Leo Desautels
2010

Digital Image Processing for Ophthalmology: Detection and Modeling of Retinal Vascular Architecture
Faraz Oloumi, Rangaraj M. Rangayyan, and Anna L. Ells

ISBN: 978-3-031-00532-9   paperback
ISBN: 978-3-031-01660-8   ebook

DOI 10.1007/978-3-031-01660-8

A Publication in the Springer series
SYNTHESIS LECTURES ON BIOMEDICAL ENGINEERING

Lecture #53
Series Editor: John D. Enderle, *University of Connecticut*
Series ISSN
Synthesis Lectures on Biomedical Engineering
Print 1930-0328   Electronic 1930-0336

Digital Image Processing for Ophthalmology

Detection and Modeling of Retinal Vascular Architecture

Faraz Oloumi
University of Calgary
Calgary, Alberta, Canada

Rangaraj M. Rangayyan
University of Calgary
Calgary, Alberta, Canada

Anna L. Ells
Alberta Children's Hospital
Calgary, Alberta, Canada

SYNTHESIS LECTURES ON BIOMEDICAL ENGINEERING #53

ABSTRACT

The monitoring of the effects of retinopathy on the visual system can be assisted by analyzing the vascular architecture of the retina. This book presents methods based on Gabor filters to detect blood vessels in fundus images of the retina. Forty images of the retina from the Digital Retinal Images for Vessel Extraction (DRIVE) database were used to evaluate the performance of the methods. The results demonstrate high efficiency in the detection of blood vessels with an area under the receiver operating characteristic curve of 0.96.

Monitoring the openness of the major temporal arcade (MTA) could facilitate improved diagnosis and optimized treatment of retinopathy. This book presents methods for the detection and modeling of the MTA, including the generalized Hough transform to detect parabolic forms. Results obtained with 40 images of the DRIVE database, compared with hand-drawn traces of the MTA, indicate a mean distance to the closest point of about 0.24 mm.

This book illustrates applications of the methods mentioned above for the analysis of the effects of proliferative diabetic retinopathy and retinopathy of prematurity on retinal vascular architecture.

KEYWORDS

computer-aided diagnosis, diabetic retinopathy, digital image processing, Gabor filters, geometrical pattern detection, Hough transform, oriented filters, parabolic modeling, pattern classification, pattern recognition, retinal fundus image, retinopathy of prematurity, shape analysis, vessel detection

Dedicated to
my love who has always encouraged and supported me to improve every aspect of
my life. This work would not have been possible without your sweet, heartening
words and your moral support.

Faraz

Contents

Preface

Retinal fundus photography is a noninvasive method of observing the human visual and nervous systems. Detection and analysis of the vasculature in fundus images of the retina can assist in the diagnosis of many different pathologies, including retinopathy, hypertension, and arteriosclerosis. A major step in performing computer-aided diagnosis (CAD) using retinal fundus images is the accurate detection of blood vessels. Other important anatomical features in the retina can be located based on the structure of vasculature. Changes to the shape, width, and structure of blood vessels can be indicative of the presence of several types of pathology, including retinopathy of prematurity (ROP) and proliferative diabetic retinopathy (PDR).

In this book, we present image processing algorithms for the detection and modeling of retinal blood vessels, including real Gabor filters for detection of the blood vessels and the generalized Hough transform (GHT) for modeling of the major temporal arcade (MTA, the thickest venular branch) in retinal fundus images. The methods can assist in terms of CAD of PDR and ROP, as shown in clinical applications.

Use of the image processing methods presented in this book is not limited to biomedical applications. Gabor filters can be used for the detection of oriented and branching patterns, and the GHT can be used for geometrical modeling of patterns and shapes. These methods can have applications in computer vision, pattern classification, and industrial nondestructive testing.

Faraz Oloumi
Rangaraj M. Rangayyan
Anna L. Ells
March 2014

Acknowledgments

We thank Dr. Fábio J. Ayres for his contributions to parts of the research work leading to the present book.

We thank the Natural Sciences and Engineering Research Council of Canada and the University of Calgary for financially supporting our projects.

We thank IEEE for permission to use parts of our paper titled "Parabolic Modeling of the Major Temporal Arcade in Retinal Fundus Images" in this book.

Faraz Oloumi
Rangaraj M. Rangayyan
Anna L. Ells
March 2014

List of Symbols and Abbreviations

$\cap$	intersection of (two sets)
$\cup$	union of (two sets)
$\subseteq$	(a subset) being entirely in (a set)
$\ominus$	Minkowski subtraction
$\in$	(an element) belongs to (a set)
$\sum$	sum of
$\delta_B(X)$	dilation of X using B
$\varepsilon_B(X)$	erosion of X using B
2D	two-dimensional
3D	three-dimensional
AMD	age-related macular degeneration
AN	actually negative
AP	actually positive
ASM	active shape model
A_z	area under the ROC curve
arctan	inverse tangent
B	a structuring element
$\check{B}$	reflection of B
CAD	computer-aided diagnosis
CFS	correlation-based feature selection
DCP	distance to the closest point
DR	diabetic retinopathy
DRIVE	Digital Retinal Images for Vessel Extraction
exp	the exponential function
FFT	fast Fourier transform
FN	false negative
FNF	false-negative fraction
FOV	field of view
FP	false positive
FPF	false-positive fraction
FT	Fourier transform

GHT	generalized Hough transform
GM	Gabor-magnitude
GUI	graphical user interface
HT	Hough transform
HVS	human visual system
IAA	inferior arcade angle
ITA	inferior temporal arcade
JPEG	Joint Photographic Experts Group
LUT	lookup table
MAT	medial-axis transformation
MDCP	mean distance to the closest point
MNN	multilayer neural network
MTA	major temporal arcade
max	the maximum value in an array or a matrix
min	the minimum value in an array or a matrix
NPDR	nonproliferative diabetic retinopathy
ONH	optic nerve head
ONHD	optic nerve head diameter
PDM	point-distribution model
PDR	proliferative diabetic retinopathy
RBFs	radial basis functions
RGB	[red, green, blue] color space
RISA	Retinal Image multiScale Analysis
ROC	receiver operating characteristics
ROI	region of interest
ROP	retinopathy of prematurity
SAA	superior arcade angle
SE	structuring element
STA	superior temporal arcade
STARE	STructured Analysis of the REtina
STD	standard deviation
TAA	temporal arcade angle
TIFF	Tagged Image File Format
TN	true negative
TNF	true-negative fraction
TP	true positive
TPF	true-positive fraction
TROPIC	Telemedicine for ROP In Calgary
VR	vertex restriction

VST   vascular skeleton tree
VTM   vascular topology map
YIQ [luminance, in-phase, quadrature] color space

CHAPTER 1

Introduction

1.1 THE HUMAN EYE

The human visual system (HVS) is the most important sensory system for gathering information, navigating, and learning. The eye is the primary sensor of the HVS, with the lens mapping the incoming light patterns onto the retina for transduction to neural signals, which are subsequently transmitted to and interpreted by the visual cortex. Several diseases have primary or secondary effects on the eye and the HVS. Therefore, examination of the eye is an important part of health care not only to assess the HVS, but also to evaluate the general well-being of the patient.

1.1.1 DIAGNOSTIC IMAGING OF THE EYE

An early method used to photograph the fundus, or the back of the eye, required the injection of fluorescein into the bloodstream to enhance the contrast of the retinal blood vessels [1]. However, over the past decade, digital fundus photography of the retina has become common practice, because it is easy to use and is noninvasive [2–4]. Digital images of the retina have been used in general eye examination, as well as to analyze and diagnose possible symptoms of certain pathologies such as diabetic retinopathy (DR), retinopathy of prematurity (ROP), arteriosclerosis, and hypertension [5–10]. Digital fundus images can be easily stored on different media without deterioration in quality and can be transmitted over short and long distances using computer networks or the Internet. Digital fundus photography can also facilitate precise comparison of the states of the retina at particular intervals of time. Another advantage of digital retinal photography is the possibility of precise measurement and monitoring of the statistics of retinal blood vessels, such as their diameter and tortuosity [11–13].

A common package of a digital fundus imaging system contains a camera, such as the CR5-NM retinal camera or the Canon Rebel XTI 8.2 MP, an imaging software system, a database management system, and a data storage device. Figure 1.1 shows a patient being examined with a digital fundus camera. Figure 1.2 (a) shows a fundus image of a normal eye from the Digital Retinal Images for Vessel Extraction (DRIVE) database [14]. For more information on images of the DRIVE database and how they were obtained, see Section 4.1.1.

1.1.2 THE ANATOMY OF THE EYE

The eye is one the most important sensors of the human body. Because the eye is a visual sensor, it produces and processes the largest amount of information among all of the human sensory systems. The eyeball, as shown in Figure 1.3, is made up of three mutually enclosing membranes

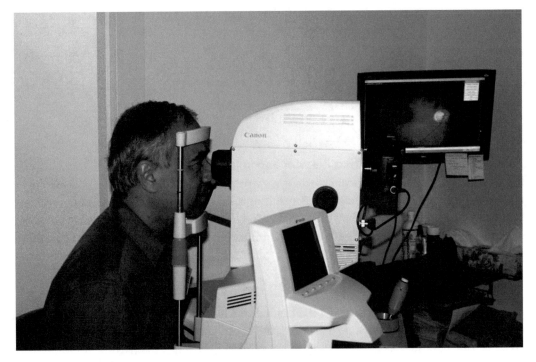

Figure 1.1: This figure shows a patient having his retina examined using a Canon digital fundus camera and the associated imaging software. The fundus image of the patient's retina is shown on a monitor at the top-right corner of the figure.

or layers: the fibrous, the vascular, and the nervous layers [15]. The fibrous layer is the outermost coat and has a protective function. The fibrous covering is mostly made up of two parts: the posterior part, called the sclera, and the anterior part, called the cornea. The sclera is a white nontransparent membrane that covers five-sixths of the globe of the eye. The cornea, on the other hand, is a transparent membrane and looks like a watch-glass. The vascular layer is the middle covering and has a nutritive function. It consists of the iris, the ciliary body, and the choroid, from front to back. The choroid is the most posterior part of the vascular layer and nourishes the outer part of the retina. The ciliary body has different functionalities, including anchoring the lens in place, changing the shape of the lens by pulling or relaxing the ciliary muscle, and providing nutrition to the avascular ocular tissues. The iris is the most anterior part of the vascular layer of the eye. The iris is similar to the diaphragm of a camera in functionality; it controls the amount of light that reaches the retina by dilating or narrowing the pupil.

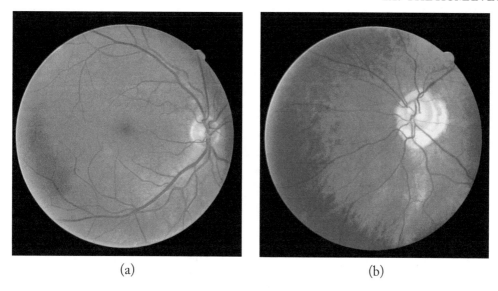

(a)   (b)

Figure 1.2: (a) Image 24 of the DRIVE database. This is a standard, macula-centered fundus image of a normal retina. Adult retinal images, such as this image, usually range from orange to brown in color. (b) Image 31 of the DRIVE database. This is a nonstandard image of the retina as it is not macula-centered. The dark regions in the image represent pigment epithelium, which can be a sign of age-related macular degeneration.

The retina is the most important part of the nervous layer of the eye. Its functionality is analogous to the film or solid-state detector in a camera; it acts as a screen onto which the focused incoming light image is projected. The photoreceptor cells in the retina, known as rods and cones, are responsible for receiving visual information from the incoming light, encoding the information in terms of neural signals, and transmitting the signals to the brain where they are interpreted as visual perception of the projected scene [16, 17]. The retina develops as an outgrowth of the forebrain during embryonic development; hence, it is the only part of the central nervous system that can be viewed directly and noninvasively [18].

The nerves that transfer the neural signals go directly from the retina to the brain through the optic nerve head (ONH), also known as the optic disc, which appears as a bright round spot in images of the retina. The average ONH diameter in adults (ONHD) is about 1.6 mm [19, 20]. The major branches of the retinal blood vessels, that is, the branches of the major vein and artery, diverge away from the trunk of the ONH into the retina. Smaller blood vessels branch off from the parent branch and converge toward a region called the macula, at the center of which is an avascular spot called the fovea. The retinal raphe is a straight line that goes through the center of the ONH and the fovea.

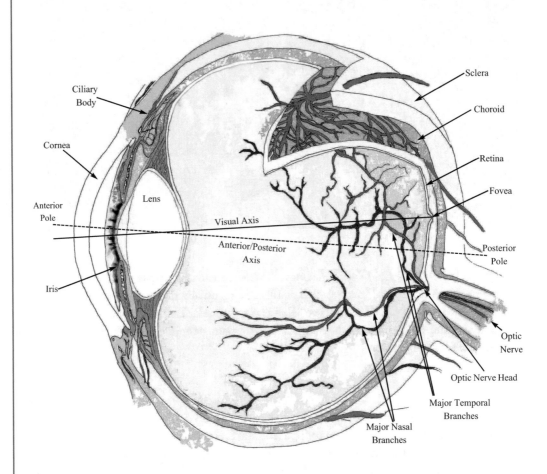

Figure 1.3: Schematic illustration of the cross section of the eyeball [15]. Different components of the membranes of the eye are labeled in this figure. The visual axis of the eye goes from the lens to the fovea as shown by a solid line. The anterior-posterior axis is shown by a dashed line. The back wall of the retina is the most posterior part of the eye and the cornea is the most anterior part. The area above the anterior-posterior axis is referred to as the temporal side, and the area below it is referred to as the nasal side of the retina. The main anatomical features of the retina are labeled in the figure. The orientation of the blood vessels in this figure do not match the orientation as in the fundus image in Figure 1.2. Standard fundus images of the retina are centered on the fovea.

Blood vessels in the retina of adults vary in thickness in the range 50–200 μm, with a median of 60 μm [5, 7]; in two-dimensional (2D) retinal fundus images, vessel thickness is defined as the edge-to-edge width of the projected image of the vessel. The capillaries are too small to be visible in typical digital retinal fundus images.

The macular region is situated approximately two ONHDs (3.2 mm) temporal to the center of the ONH [19] along the retinal raphe. The area above the retinal raphe is called the superior side, and the area below the retinal raphe is called the inferior side. The macula appears darker than the surrounding background and has the highest density of photoreceptors in the retina. The fovea appears as a small bright spot approximately in the middle of the macula and is situated at the end of the visual axis (see Figure 1.3). The major venule branch is thicker than the corresponding arteriole branch and has a higher background contrast; it is named the major temporal arcade (MTA) because of its arch-like structure and proximity to the temple. Figure 1.4 shows a typical fundus image of the retina from the DRIVE database, with the main anatomical features annotated.

As shown in Figure 1.3, the portion of the retina that is below the posterior pole is called the nasal side, and the portion that is above the posterior pole is called the temporal side. Most of the standard images of the retina are macula-centered and show the temporal side of the retina, so that all of the important diagnostic features of the retina are visible. Figure 1.2(a) shows a standard macula-centered image. However, some retinal images are ONH-centered; hence, they may not display the macula or all parts of the MTA. Figure 1.2(b) is an example of an ONH-centered image of the retina; the macula and substantial parts of the MTA are not visible in this image.

1.2   PATHOLOGIES OF THE EYE

As mentioned in Section 1.1.1, several abnormal conditions, such as hypertension, arteriosclerosis, DR, and ROP can affect the structure and anatomical features of the retina. In general, most of the research on image processing applied to fundus images of the retina has been focused on the detection and analysis of the changes that occur due to DR and ROP. The following sections provide details on the symptoms of DR and ROP, and give a brief overview of some of the other diseases that can affect the retina.

1.2.1   DIABETIC RETINOPATHY

DR has been the leading cause of preventable blindness among people of working age in developed countries [5, 21–23]. DR is commonly divided into two main categories of nonproliferative DR (NPDR) and proliferative DR (PDR). NPDR can be characterized in terms of several types of lesions, such as microaneurysms, cotton-wool spots, hemorrhages, drusen, and exudates; it can also be diagnosed by the presence of macular edema and venous beading [5, 23, 24]. The presence of these features has been shown to correlate with the severity and progression of NPDR [25, 26]. Figure 1.5 shows a retinal fundus image from the STructured Analysis of the REtina (STARE) database that shows signs of NPDR in terms of exudates. One of the signs that is indicative of

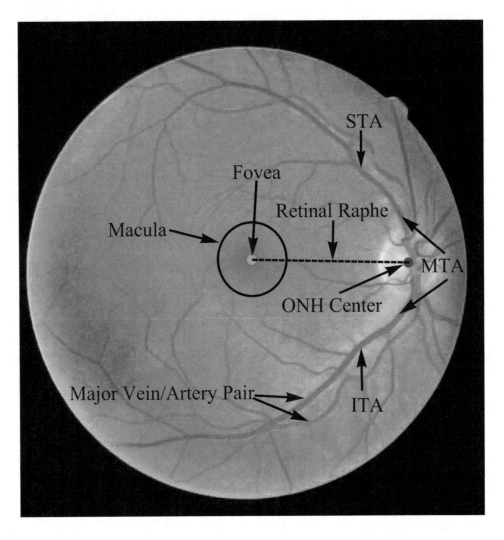

Figure 1.4: Image 24 of the DRIVE database. The main anatomical features are annotated. The circle near the middle of the image approximates the macular region. The green dot in the middle of the circle indicates the fovea. The center of the ONH, the point of divergence of the blood vessels, is marked by a blue dot on the right-hand side of the image. The straight line that goes through the center of the ONH and the fovea is the retinal raphe. The area above the retinal raphe is the superior side and the area below it is the inferior side. The MTA consists of two parts: the inferior temporal arcade (ITA) and the superior temporal arcade (STA). The ITA and the STA are the top and bottom parts of the major venule, respectively. The inferior major vein-artery pair is also labeled.

the presence of PDR is characterized in terms of changes to the architecture of the MTA, due to proliferation of fibrovascular tissue over the surface of the retina [27, 28]. Contraction of the surface of the retina, as well as the fibrovascular proliferation of the pathological tissue near the ONH, cause the MTA to be pulled toward the periphery of the retina, resulting in tractional retinal detachment and subsequent loss of vision [27, 28].

The detection of the main anatomical features of the retina, described in Section 1.1.2, is an important preprocessing step for the identification and analysis of the signs of NPDR and PDR. Lesions that are indicative of NPDR usually appear within a circle of radius one ONHD from the fovea [29]; hence, it would be beneficial to analyze only the macular region for this purpose. The main anatomical features, such as the blood vessels and the ONH, may be removed, after being detected, to reduce the chance of false detection of lesions. There have been many studies on automatic detection and diagnosis of NPDR using image processing techniques, some of which are discussed in Section 1.3. Detection, modeling, and analysis of the openness of the MTA could also help in the diagnosis of PDR, as demonstrated in Chapter 7.

1.2.2 RETINOPATHY OF PREMATURITY

Plus Disease

ROP is the leading cause of childhood blindness, which is preventable by laser surgery only if it is diagnosed at its early stages [30, 31]. The severity of ROP can be categorized by the presence of plus disease [30, 32, 33]. Plus disease has been difficult to define in a quantitative manner; diagnosis is usually made by visual qualitative comparison to a standard photograph of the fundus of the retina [30, 32, 34, 35]. If plus disease is not diagnosed early, ROP can advance to its late stages, and the outcome of treatment may not be favorable [31]. Even though plus disease is graded simply as being present or absent, experts in the field agree that there is a range of associated posterior vascular changes [30, 33–35]. Changes that are indicative of the presence of plus disease include abnormal dilation and tortuosity of blood vessels near the posterior pole, as well as a decrease in the angle of insertion of the MTA [5, 24, 30, 32, 36, 37].

Interobserver Variability in the Diagnosis of Plus Disease

As Freedman et al. [38] observed, even nonexperts can detect small changes in dilation and tortuosity of the temporal vessels by comparing a given image to a standard image showing the minimum dilation and tortuosity due to plus disease. However, in a study set up to measure the level of agreement between three experienced ROP examiners in diagnosing plus disease, there was disagreement in 27% (18 of 67) of the images judged to have plus disease [39]. Wallace et al. [39] also reported 37% (25 of 67) and 31% (21 of 67) disagreement on the presence of sufficient tortuosity and dilation, respectively, that is indicative of plus disease. Such studies and the importance of positive diagnosis of plus disease indicate the need for computer algorithms to detect retinal vasculature and analyze the associated patterns to quantify different diagnostic factors, such as vessel thickness and tortuosity. Another diagnostic factor, which has been quan-

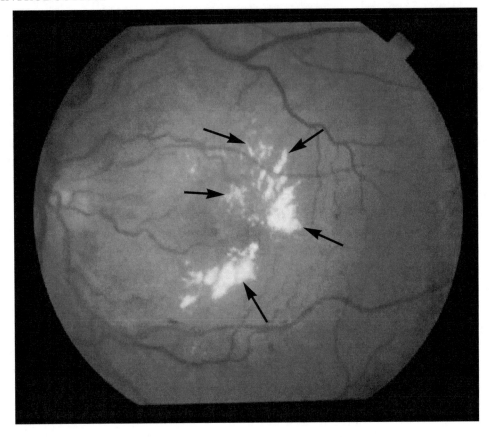

Figure 1.5: Image 0001 of the STARE database, which shows signs of NPDR, manifested as exudates (the bright lesions pointed to by arrows).

tified manually in only a very few studies [40, 41], is the decrease in the angle of insertion of the MTA [30, 33, 42]: this is the main focus of the methods described in this book.

Changes in Vessel Thickness and Tortuosity
Many algorithms and programs have been developed to detect and quantify changes in vessel thickness and tortuosity, and to correlate them with the presence of plus disease [7, 13, 43–45]. Heneghan et al. [43] reported an average vessel thickness and tortuosity increase of 9.6 μm and 0.008 for infants who progressed to threshold disease (a level of progressive ROP that warrants treatment) in a screening program; only the change in width was found to be statistically significant. Using the Retinal Image multiScale Analysis (RISA) program, Swanson et al. [7] reported

median differences of 7.4 μm, 6.5 μm, and 0.092 in venule diameter, arteriole diameter, and arteriole tortuosity, respectively, between cases with no ROP and severe ROP. Arteriole tortuosity was found to vary significantly with the severity of ROP. However, the changes in the diameter of venules and arterioles were found to have no statistical significance. Also using RISA, Gelman et al. [35] reported median diameter differences of 12.8 μm for arterioles and 12.1 μm for venules due to plus disease. They also found the changes in arteriole tortuosity to be more consistent with the presence or absence of plus disease. By using the Computer Assisted Image Analysis of the Retina program, Wilson et al. [32] reported that changes in the diameter of venules are difficult to detect as they are at, or below, the limit of spatial resolution of typical pediatric retinal fundus images.

Even though a direct relationship seems to exist between increasing venule thickness and the presence of plus disease, it has not been shown to have any statistical significance. Also, the detection of small changes in thickness requires high-resolution imaging. Arteriole tortuosity has shown higher correlation with the presence of plus disease, but this has not been consistent across all trials [43]. Also, the detection of arterioles presents an image processing challenge, because their diameters are smaller than those of venules, they have lower background contrast, and their thickness can be below the resolution limit of even high-resolution images in preterm infants [7]. These factors and limitations indicate a great need for quantification of the third diagnostic parameter in plus disease: the angle of insertion of the MTA.

Changes in the Angle of Insertion of the MTA

The angle of insertion of the MTA has been loosely defined as the angle between the STA and the ITA as they diverge from the ONH and extend toward the periphery of the retina [40, 41]. This angle, also called the temporal arcade angle (TAA), is used as an indicator of the structural integrity of the macular region [24, 40, 41].

A decrease in the TAA has been cited as a manifestation of at least two types of pathology: as a sequela of ROP [5, 24, 30, 32, 36], and as the severity indicator of myopia [41, 46]. As previously mentioned, the architecture of the MTA is also known to be affected by PDR due to tractional retinal detachment [47] (see Section 1.2.1). Despite the clinical importance of abnormal values of the TAA, it has been quantified in only a few studies: a study dealing with myopia [41] and two studies dealing with ROP [40, 48] have analyzed manual measurements of the TAA and correlated changes in the TAA with the stage of the diseases mentioned.

Fledelius and Goldschmidt [41] measured the TAA and correlated its decrease with progression of myopia based on follow-up data over a 38-year period. They defined the TAA by manually marking cardinal points at the first or the second arteriole-venule crossings (decided subjectively to best represent the direction of the MTA) away from the ONH, with the vertex of the angle being the center of the ONH. The cardinal points were used as landmarks from image to image. Two lines were drawn from the center of the ONH to the marked cardinal points on the ITA and the STA. The angle between the two lines was measured using a transparent angle meter. Fledelius and Goldschmidt [41] reported a decrease of more than 4° in the TAA in 25% (6 of 24) of the cases with high and stable myopia and in 60% (12 of 20) of the cases with high and progressive myopia. The change in the TAA of the progressive myopia group, as compared to the stable myopia group was shown to be statistically highly significant ($p < 0.01$). For the high and progressive myopia group, the change in the TAA was shown to be correlated with the degree and increase of myopia.

Change in the TAA also has been featured in the classification of retrolental fibroplasia [42], and, more recently, in the classification of ROP [30]; it also has been used in the evaluation of structural changes following cryotherapy [31]. The Cryotherapy of Retinopathy of Prematurity Cooperative Group [31] evaluated the TAA by tracking the MTA in 30° sectors on photographic images; however, the normal range of the TAA was not defined. By combining time series of color fundus images into short video clips, Ells and MacKeen [49] illustrated that the changes that occur in the MTA in the presence of progressive ROP are dynamic, as they alter the architecture of the MTA.

Wilson et al. [40] defined the TAA as follows: the center of the ONH and the fovea are manually marked. A line is drawn through the marked center of the ONH and the fovea; this is the retinal raphe. The image is rotated so that the retinal raphe is horizontal, as shown in Figure 1.6. A line perpendicular to the retinal raphe is drawn from the fovea until it intersects the ITA and the STA. From these intersections, two lines are drawn to the center of the ONH. The angle between the retinal raphe and the line connecting the center of the ONH to the STA is the superior arcade angle (SAA). The angle between the line connecting the center of the ONH to the ITA and the retinal raphe is the inferior arcade angle (IAA). The TAA is defined as the sum of the SAA and IAA, as shown in Figure 1.7.

Wilson et al. [40] reported a high degree of interocular symmetry with a mean TAA of 82° for both eyes. They indicated that interocular asymmetry of above 14° to 20° between the two eyes of a patient should be treated with suspicion. They associated a significant level of angle acuteness in the left IAA between stage 0 and 1, stage 1 and 2, and stage 1 and 3 ROP (higher numbers indicate higher severity of ROP).

In a related follow-up study by Wong et al. [48], semiautomated measurements were made of four different angles of the temporal and the nasal venules and arterioles. The procedures required manual editing of automatically detected vessels; this step required 10–15 minutes per image. As compared to the previous related study of Wilson et al. [40], the angles were measured using reference points selected closer to the center of the ONH. The nasal angles were found to have no statistically significant differences between normal cases and ROP of various stages. The angles of the temporal venules and arterioles were found to have statistically significant differences between normal cases and stage 3 ROP. However, when all stages of ROP were combined, only the angle of the temporal arterioles indicated a statistically significant difference as compared to the normal cases.

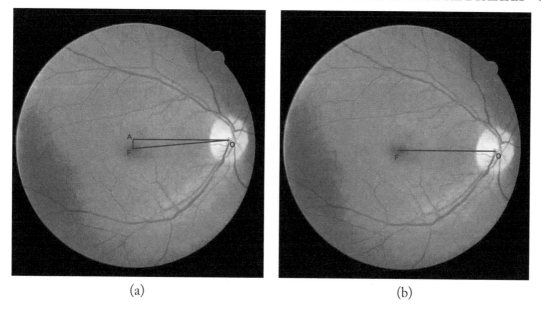

(a)   (b)

Figure 1.6: (a) Image 27 of the DRIVE database showing the manually marked fovea (F) and the center of the ONH (O) (see Section 4.2 for the details of the annotation process). The line OF represents the retinal raphe. The retinal raphe angle is defined as $\theta = \arctan(\frac{AF}{OA})$. (b) The image in (a) with the raphe angle corrected, i.e., $\theta = 0°$.

1.2.3   OTHER RETINAL DISEASES

The diseases that affect the retina are not limited to DR and ROP; there are several other pathologies that can be diagnosed by analyzing retinal images. These include, but are not limited to, glaucoma [50], age-related macular degeneration (AMD) [51], hypertension [52], and cardiovascular diseases [53].

1.3   COMPUTER-AIDED DIAGNOSIS OF RETINAL DISEASES

Screening programs have been implemented at health-care centers around the world to facilitate early diagnosis of retinal diseases [5, 54, 55]. Such programs require an ophthalmologist to examine large numbers of retinal fundus images, search for possible abnormalities, and provide the diagnostic results accordingly. Computer-aided diagnosis (CAD) via analysis of retinal images by image processing techniques and pattern analysis methods could provide a number of benefits. CAD can reduce the workload and provide objective decision-making tools to ophthalmologists.

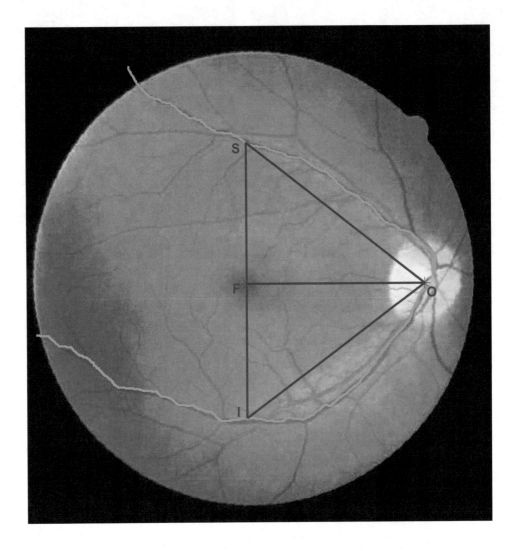

Figure 1.7: The same image as in Figure 1.6 (b) illustrating the procedure for measuring the TAA. The MTA, as traced by a retinal specialist (see Section 4.2), is highlighted in green. Points 'S' and 'I' represent the points of intersection of the normal (SI) to the retinal raphe (OF) with the STA and ITA, respectively. The angle $\angle SOF$ is the SAA, the angle $\angle IOF$ is the IAA, and the angle $\angle SOI$ is the TAA. Hence, TAA = SAA + IAA, where SAA $= \arctan(\frac{SF}{OF})$ and IAA $= \arctan(\frac{IF}{OF})$.

Computerized algorithms can also facilitate precise measurement of different parameters and can quantify small changes in the measurements [5, 39].

It has been shown that automated detection of the pathologies mentioned in Section 1.2 is possible, and can lead to early treatment [24]. CAD requires automated segmentation and detection of the main features of the retina, such as blood vessels, the ONH, and the macula. The positioning of the main anatomical features can help localize the positions of lesions, which then can be detected automatically [5, 9, 56–67]; the detection and localization of lesions could help in diagnosing AMD, NPDR, and other pathological conditions. Analyzing changes in the shape and color of the ONH can assist in detecting the presence of glaucoma [68–70]. Analyzing changes in the statistics and architecture of blood vessels can assist in the diagnosis of plus disease [37], PDR[47], and myopia [41].

As mentioned in Section 1.2.2, there are several studies that have correlated the severity of ROP to changes in thickness and tortuosity of retinal blood vessels using image processing programs [7, 13, 43–45]. However, the studies that correlated changes in the TAA to the severity of ROP were based on manual input and no CAD system has been designed for this purpose.

1.4   SCOPE AND ORGANIZATION OF THE BOOK

The main aim of the work underlying the present book is the development of digital image processing and pattern analysis techniques, first to detect the blood vessels, and second, to obtain a parametric model of the MTA, in fundus images of the retina. The contents of this book are organized in seven chapters and a list of references.

Chapter 2 presents a brief overview of image processing methods for the detection of the main anatomical features of the retina in fundus images.

Chapter 3 presents an overview of image processing techniques that have been implemented to address the stated problems, including morphological operators, detection of oriented structures, and detection of geometrical patterns.

Chapter 4 provides details of the database used in the present work and the methods used for the evaluation of the results of detection of blood vessels and modeling of the MTA.

Chapter 5 presents detailed descriptions of the procedures for the detection of blood vessels using single-scale, multiscale, and multifeature analyses; it also presents an analysis of the obtained results.

Chapter 6 provides detailed descriptions of the procedures for parametric modeling of the MTA and presents an analysis of the results obtained.

Chapter 7 presents a discussion on potential clinical applications of the methods described in the preceding chapters, as well as notes on future work on computer-aided analysis of images of the retina.

CHAPTER 2

Computer-aided Analysis of Images of the Retina

2.1 DETECTION OF ANATOMICAL FEATURES OF THE RETINA

CAD of diseases affecting the eye, as mentioned in Section 1.3, requires the preliminary detection and delineation of the normal anatomical features of the retina, including the blood vessels, the macular region, and the ONH [71–73]. The detection of several anatomical features can help in screening for certain diseases by analyzing the pathophysiological changes that occur in such features, as mentioned in Section 1.2. The locations of the main anatomical features can also be taken as references or landmarks in the image for the detection of pathological features, such as microaneurysms, hemorrhages, exudates, macular edema, venous beading, and neovascularization [5]. It may also be useful to mask out the normal anatomical features in a given image in order to detect only the pathological features.

2.1.1 DETECTION OF BLOOD VESSELS

In many applications of image processing in ophthalmology, the most important step is to detect the blood vessels in the retina [5, 9, 71, 74–80]. Furthermore, the location of certain features, such as the ONH and the macula, can be found relative to the vascular architecture or the MTA [72, 74, 78, 81–83]. This section presents a review of some of the previously proposed methods and algorithms for the detection of blood vessels in the retina.

Matched filters: Chaudhuri et al. [71] proposed an algorithm based on 2D matched filters and three assumptions: vessels can be approximated by piecewise linear segments, the intensity profile of a vessel can be approximated by a Gaussian curve, and the width of vessels stays constant. Detection is performed by convolving the given image with the matched filter rotated in several directions, with the maximum response recorded for each pixel.

Adaptive local thresholding: The method of adaptive local thresholding using a verification-based multithreshold probing scheme was used by Jiang and Mojon [84]. In this method, a binary image obtained after applying a threshold is used in a classification procedure to accept or reject any region in the image as a certain object. A series of different thresholds is applied, and the final detection result is a combination of the results provided by the individual thresholds.

Ridge-based vessel segmentation: The assumption that vessels are elongated structures is the basis for the supervised method of ridge-based vessel detection and segmentation, which was introduced by Staal et al. [77]. The ridges in the image, which roughly coincide with the vessel center-lines, are extracted by this algorithm. Then, image primitives are obtained by grouping image ridges into sets that model straight-line elements. Such sets are used to partition the image by assigning each pixel to the closest primitive set. In each partition, a local coordinate is defined by the corresponding line element. Finally, feature vectors are computed for every pixel using the characteristics of the partitions and their line elements, and classified using sequential forward feature selection and a k-nearest-neighbor classifier. Staal et al. achieved an area (A_z) under the receiver operating characteristic (ROC) curve of 0.9520 for the 20 images of the test set of the DRIVE database (see Section 4.1.1 for details regarding the DRIVE database and Section 4.3.1 for a description of ROC analysis).

Piecewise threshold probing of a matched-filter response: This method, proposed by Hoover et al. [76], uses local vessel attributes as well as global and region-based attributes of the vascular network for the detection and classification of vessels. Different areas and regions in a matched-filter response are probed at several decreasing thresholds. At each level, the region-based attributes are used to determine whether the probing should be continued, and to classify the probed area as a blood vessel or not.

Vessel segmentation using 2D complex Gabor filters and supervised classification: An algorithm applying complex Gabor filters (see Section 3.2.2) for feature detection and supervised classification of blood vessels was proposed by Soares et al. [85]. In this method, a feature vector containing the magnitude output at several scales, obtained from 2D complex Gabor filters, is assigned to each pixel. In the next step, using a Bayesian classifier with class-conditional probability density functions given by a Gaussian mixture model, each pixel is classified as a vessel or nonvessel pixel. Soares et al. reported $A_z = 0.9614$ for the 20 test images of the DRIVE database.

Multiscale vessel segmentation using real Gabor filters: Rangayyan et al. used real Gabor filters at multiple scales of thickness for the detection of the blood vessel in retinal images [86]. Radial basis functions (RBFs) were used to perform multiscale analysis on the results of Gabor filtering. An area under the ROC curve of $A_z = 0.96$ was obtained for the 20 test images of the DRIVE database. See Section 3.2.2 for details of real Gabor filters.

Active contour model: Al-Diri et al. [87] used the ribbon-of-twins model, which consists of a pair of twin contours (dual and sandwich snakes), to detect the edges of blood vessels. A tramline algorithm is used in this method to identify potential vessel center-line pixels by employing a generalized morphological filter similar to the grayscale top-hat filter. Even though the tramline algorithm is expected to detect only the possible center-line pixels, the morphological process of skeletonization is used on the result of the tramline algorithm to identify further the approximate center lines. The tramline pixel map is used as the input to the segmentation algorithm, which uses the twin-ribbons algorithm to grow the vessel regions; the two ribbons converge from the outside and the inside of the vessel toward the edges. Finally, a procedure is used to connect the segments at junctions that are left unconnected by the segment-growing algorithm. To obtain a valid vascular skeleton tree (VST), any pixel whose center point lies inside a segment (between the two edge segments) is regarded as a vessel pixel. The segmentation performance of the proposed algorithm was obtained in terms of sensitivity and specificity by comparing the results against the ground-truth images of the DRIVE database. A sensitivity of 72.82% was achieved at 95.51% specificity (see Section 4.3.1 for definitions of sensitivity and specificity).

Contour evolution using color components: By considering the local linearity, piecewise connectivity, and brightness of vessels, Ushizima et al. [88] used a Gaussian-like profile to enhance vessel-like patterns in the green-channel components of the DRIVE images. The morphological process of top-hat was used to approximate roughly a vessel segment using a straight line with a length 15 pixels as the structuring element (SE). The SE was rotated in 15° increments. The output of the top-hat process, I_{th}, was thresholded to provide seeds (belonging to vessels) to initialize the front-evolution algorithm. I_{th} was also taken as the speed function used in the evolution procedure along with the information provided by color components in the vicinity of the propagation front. Ushizima et al. proposed a fast-marching algorithm, which propagates only if the color similarity between the current position and the new position is guaranteed. A sensitivity of 87% at 85.6% specificity was reported.

Multifeature supervised vessel segmentation: Marin et al. [89] obtained seven features, including grayscale and moment-invariant-based features in order to classify a pixel. The preprocessing steps used include removal of the central light reflex of the vessels, homogenization of the background, and enhancement of the appearance of the vessels. Feature extraction was achieved using the vessel-enhanced image. Five grayscale features, based on the differences between the grayscale value of a given pixel and a statistical representation of its surroundings in terms of the mean, standard deviation (STD), and other statistical measures, were calculated. Two features were also computed using 2D moments of different orders. A neural network was trained using a small set of pixels, selected manually, and their corresponding ground-truth values. A postprocessing step was used to improve further the classification result of the neural network by filling in holes in the vessels and by removing unwanted small groups of pixels. Marin et al. reported an area under the ROC curve of $A_z = 0.9588$ for the 20 test images of the DRIVE database.

The vesselness measure: Frangi et al. proposed a method for the detection of vessel-like patterns by analyzing the properties of the eigenvalues of the Hessian matrix, which was computed at multiple scales by convolving a given image with Gaussian kernels of different scales [90]. In a recent study [91], it was shown that the vesselness measure does not perform as well as the real Gabor filters in terms of accuracy in the detection of blood vessels in retinal images based on the results of ROC analysis.

Lupaşcu et al. [92] combined the features obtained using some of the previously mentioned methods with features that represent information on the local intensity and structure of vessels, as well as information on the spatial properties and geometry of the vessels at different scales of length. Eight different types of features were obtained:

- region- and boundary-related features in Gaussian scale space [93],

- model-based vessel likelihood [94],

- Frangi vesselness [90],

- Lindeberg ridges [93],

- Staal ridges [77],

- complex Gabor filter magnitude response [85],

- responses of differential second-order detectors for elongated structures, and

- intensity-based vessel likelihood.

Most of the features listed above were obtained at multiple scales or with variations of a defining parameter, with the resulting feature vector containing 41 features. An iterative boosting algorithm, called AdaBoost [95], was used to train a classifier using the training set of the DRIVE database. At first, all 41 features were used in training the classifier, which was then applied to the test set. The authors reported an area under the ROC curve of $A_z = 0.9561$.

Lupaşcu et al. also applied various feature selection methods to the 41-dimensional feature vector in order to determine which features may have more impact on the training procedure, and subsequently, on correct classification of the vessels [96]. Five feature selection methods were applied, all of which were correlation-based feature selection (CFS) methods: CFS using hill climbing search, CFS using best first search, consistency-based CFS, entropy-based CFS, and CFS using t-statistics. The five methods selected 7, 4, 21, 20, and 16 features, respectively. The AdaBoost classifier [95] was trained using data only from the selected features and ROC analysis was used again to determine the classification accuracy. However, none of the combinations of features led to a higher A_z value, as compared to the results achieved using all 41 features ($A_z = 0.9561$).

Other methods reported for the detection of blood vessels in the retina include segmentation using multiconcavity modeling [97]; detection using a modified matched filter [98]; the use of amplitude-modified second-order Gaussian filters [99]; vessel models and the Hough transform (HT) [100]; the Gabor variance filter with a modified histogram equalization technique [101]; mathematical morphology and curvature evaluation [102]; and tramline filtering [103]. Several techniques have been proposed to model and analyze the structure of retinal vasculature, including fractals [104, 105] and geometrical models and analysis of topological properties of the blood vessels [74, 75, 106].

2.1.2   DETECTION OF THE ONH AND THE MACULA

The ONH appears as an approximately circular area brighter than its surrounding region, and is the point of divergence of the network of blood vessels [1]. Changes in the properties of the ONH can indicate the presence of certain pathologies (see Section 1.2). Blood vessels diverge toward the macular region, which is usually the darkest spot in retinal fundus images [1]. Changes in the measured area of the avascular macular region can indicate the severity of DR [107].

Several methods have been proposed to detect the ONH; these methods include using the shape and color properties of the ONH [108–112], matched filters [20, 113, 114], geometric models [72, 74], a fractal-based method [115], and using the divergence point of the blood vessels [73, 116].

The macular region can be located after the detection of the ONH, as it is situated approximately two ONHDs temporal to the ONH center [19] along the retinal raphe [78, 81]. The macula also can be detected by analyzing the overall structure of the VST, because the blood vessels diverge away from the ONH toward the macular region [83, 109, 117].

2.1.3 DETECTION OF THE MTA

Because the MTA originates from the ONH and follows a curved, almost parabolic, path toward the macula, it can be used to detect or estimate the position of the ONH. Furthermore, relative to the location of the ONH, the macular region also can be estimated [72, 74, 78, 81–83]. The following paragraphs give a brief overview of the methods used to detect and locate the MTA.

Foracchia et al. [74] proposed a method for the detection of the ONH by defining a directional model for the vessels, assuming that the main vessels (parts of the MTA) originate from the ONH and extend in paths that can be geometrically modeled as parabolas. A directional model was defined using the parabolic formulation and assuming that the preferred direction of the vessels is tangential to the parabolas themselves. With the model and data indicating the center points, direction, and caliber of the vessels, and by using a residual sum-of-squares method, the parameters of the model were identified.

Using steerable filters and color thresholding, Kochner et al. [82] extracted edge points on the main blood vessels. An ellipse was then fitted to these points using the generalized Hough transform (GHT). The end of the long axis of the ellipse was taken as an estimate of the location of the ONH.

Using an active shape model (ASM) and by defining a point-distribution model (PDM), Li and Chutatape [78] proposed a method to detect the boundary of the ONH and the main course of the blood vessels. Using the ASM and principal component analysis, the location of the ONH was estimated. A modified ASM was used to extract the main course of the blood vessels. Thirty landmark points on the main course of the vessels were used to derive the PDM. The HT and linear least-squares fitting methods were combined to estimate a parabolic model.

Using an estimate of the ONH location and a binarized image of the vasculature, Tobin et al. [81] proposed to apply a parabolic model to the statistical distribution of a set of points given by a morphologically skeletonized vascular image to find an estimate of the retinal raphe. A parabola of the form $ay^2 = |x|$ was modified to accommodate for the shifted vertex at the most likely ONH location and the angle of rotation of the retinal raphe, β. The resulting model and the skeletonized image were used with a least-squares method to estimate the parameters a and β. Even though Tobin et al. estimated the openness of the parabolic model, it was only used to draw a parabola on the image and was not used to quantify the openness of the MTA.

Fleming et al. [72] proposed a method to extract the MTA by means of vessel enhancement and semielliptical curve fitting using the GHT. First, the vessels were enhanced to get a magnitude image and a phase image of the vascular architecture. Having an edge map and knowing the orientation of the arcade, a reference point can only be at one of a few locations, the GHT was applied to a skeletonized image of the vasculature. The dimension of the Hough space was set to be five, with variables for inclination, horizontal axis length, left or right opening, and the location of the center of the semiellipse. Anatomical restrictions were applied to the variables to limit the number of semiellipses generated by the method. The global maximum in the Hough space was selected as the closest fit to the MTA.

Ying and Liu [83] obtained a vascular topology map (VTM) using an energy function defined as the normalized product of the local blood vessel width and density. A quantile threshold was used on the VTM to extract the pixels in a high-energy band. A circle-fitting method was applied to the extracted pixels to model the MTA as a circle, which was then used to localize the macula.

In a previous study [118], we used single- and dual-parabolic modeling via the GHT to model and quantify the openness of the MTA, STA, and ITA in retinal images. Chapter 6 provides the details and results of the mentioned methods.

2.2   LONGITUDINAL ANALYSIS OF IMAGES OF THE RETINA

Methods for longitudinal analysis of images of the retina are designed to facilitate the follow-up of patients with retinal pathology and to perform an assessment of the effects of treatment [41, 119, 120]. Fledelius and Goldschmidt [41] measured the TAA and correlated its decrease to the changes in the degree of severity of myopia based on follow-up data over a 38-year period. Ells et al. [119] evaluated the use of remote reading of digital retinal photographs in the diagnosis of severe ROP in a program for longitudinal screening for ROP. Narasimha-Iyer et al. [120] proposed algorithms to detect and classify changes in time series of color fundus images of the retina.

2.3   REMARKS

A general review of computer algorithms for the detection of anatomical features of fundus images of the retina was presented in this chapter. The features of interest included the blood vessels, the ONH, the macula, and the MTA. Algorithms for the detection of pathological features in retinal images have not been reviewed in this chapter. The methods developed in the course of the research work leading to the present book for the detection of retinal blood vessels and the

MTA are described in detail in Chapter 3. The performance of Gabor filters in the detection of retinal blood vessels is evaluated in comparison with some of the works reviewed in the present chapter, where the same database (DRIVE) was used and the methods were evaluated using ROC analysis.

Even though the structure of the MTA has been used to estimate the location of the ONH and the macula in previously reported works, only Tobin et al. [81] modeled the arcade for parameterization of its openness. However, they used the openness parameter only to draw the parabolic model on the image; hence, none of the reviewed methods has provided a performance measure regarding the accuracy of the algorithms used for the detection of the MTA, or used the result in a diagnostic application. Methods for quantification of the performance of an MTA modeling procedure are described in Section 4.3 and analyzed in Section 6.8. Potential clinical applications are demonstrated in Chapter 7.

CHAPTER 3

Methods for the Detection of Oriented and Geometrical Patterns

Methods to address the problem of detection of retinal blood vessels may take advantage of the fact that blood vessels are elongated, piecewise-linear, or curvilinear structures with a preferred orientation [71, 77, 85, 98, 101]. One of the widely used methods for the detection of blood vessels is by means of matched filtering [71, 76, 77, 98]; a general overview of matched filters is given in Section 3.2. Real Gabor filters, which are oriented-feature detectors, can be applied to images of the retina for the detection of blood vessels, as described in detail in Section 3.2.2. The angle of orientation of the blood vessels, as provided by the Gabor filters, can be used to compute coherence to provide an additional measure for multifeature analysis for the detection of the blood vessels, as explained in Section 3.2.3. Detection of the retinal blood vessels is a necessary step prior to modeling of the MTA. The known geometrical shape of the MTA can be utilized in the modeling procedure. The HT, which is an image processing method for the detection of geometric patterns and parametric curves, is used to model the MTA, as explained in Section 3.3. Binary morphological operators, which can be used to modify and filter binary images in the blood-vessel detection and the MTA modeling procedures, are described in detail in Section 3.1. Pattern classification methods used to perform multiscale and multifeature analysis in the detection of blood vessels are briefly described in Section 3.4.

3.1 MORPHOLOGICAL OPERATORS

Mathematical morphology, which is based on set theory [121], forms the foundation of morphological image processing techniques [122]. Set theory deals with the interactions of different sets, which can be represented as different objects in a binary image. In binary morphological image processing, there are usually two sets involved; one set is considered to represent the objects in the binary image (pixels that have a value of one against a background of zero-valued pixels), and the other set is called a structuring element (SE), which is of a predefined shape and geometry [122]. The shape of the SE is determined by the relevant information that needs to be processed in the image. An SE is essentially a small set of pixels, where the shape of the SE is represented with pixels that have a value of one against a background of zero. Figure 3.1 shows a few common SE shapes; these binary SEs are said to be flat. SEs that are applicable to grayscale images are said to be nonflat or volumic [123]. An SE, B, is said to be symmetric if it is identical to its reflection, $\check{B}$, when reflected about its origin [123].

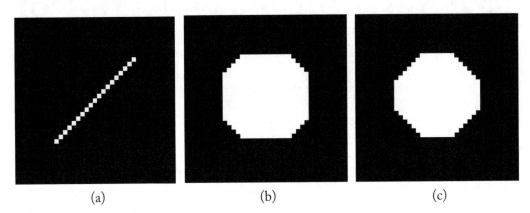

|(a)|(b)|(c)|

Figure 3.1: Common SE shapes: (a) a line of length $L = 19$ pixels oriented at $\theta = 45°$, (b) a disk of radius $r = 10$ pixels, and (c) an octagon with $R = 9$ pixels, where R is the distance from the structuring element's origin to the vertical and horizontal sides of the octagon. The SEs shown in (b) and (c) are symmetric. All three SEs are created using a 19×19 matrix; the shape of the SE is represented by the pixels that have a value of one in this matrix. For the sake of illustration, the SEs are superimposed on a black background of size 38×38 pixels.

3.1.1   BINARY EROSION

Binary erosion is the most basic morphological operator and can be represented in terms of the intersection of two sets. Given a set, X, and an SE, B, the erosion of X using B is defined as [123, 124]

$$\varepsilon_B(X) = \bigcap_{b \in B} X_{-b}, \tag{3.1}$$

where b determines the translation of (shift in) B and the minus sign implies a reflection. The result of erosion can be thought of as a set of points, x, where B will fit entirely inside X, if the origin of B is placed at x. In mathematical terms,

$$\varepsilon_B(X) = \{x \mid B_x \subseteq X\}, \tag{3.2}$$

where B_x denotes the SE centered at x. For practical implementation, binary erosion can be formulated as

$$\varepsilon_B[f(x)] = \min_{b \in B} f(x + b), \tag{3.3}$$

where f is the given image. Equation 3.3 implies that erosion changes the value of a given pixel x, to be equal to the minimum value of the image f in the area defined by the SE, when B is centered at x. In other words, the pixel where B is centered will be eroded (turned off) if at least one pixel from the binary image, f, is equal to zero in the window defined by B. The binary erosion of a set, X, using an SE, B, is equivalent to the Minkowski subtraction of B from X if and only if the SE is symmetric ($B = \check{B}$). Minkowski subtraction is given as [123]

$$X \ominus B = \bigcap_{b \in B} X_b. \tag{3.4}$$

Binary erosion is given as

$$\varepsilon_B(X) = \bigcap_{b \in B} X_{-b} = \bigcap_{-b \in B} X_b = \bigcap_{b \in \check{B}} X_b = X \ominus \check{B}.$$

However, if $B = \check{B}$, then $\varepsilon_B(X) = X \ominus B$.

Figure 3.2 shows the effect of binary erosion on a test image using different SEs. The images in Figure 3.2 are shown in color for better representation of the erosion process. However, the actual test image used in this example is a binary image, where the shapes are white. Note that a square SE with side $s = 10$ pixels has a larger area than a disk-shaped SE with diameter $d = 10$ pixels; hence, a larger area will be eroded using a square-shaped SE as compared to that using a disk-shaped SE; this is evident by comparison of the green areas in Figure 3.2 (d) to those in part (c) of the same figure.

In the present work, binary erosion is used in the process of generation of masks for the images of the DRIVE database to erode the edge artifacts caused by thresholding a grayscale image to a binary image. See Section 4.1.1 for information on the images of the DRIVE database. Refer to Section 5.2 for details of the mask-generation process.

3.1.2 BINARY DILATION

Binary dilation is, in theory, similar to binary erosion, as they follow the duality property with respect to complementation; binary dilation of an image is the complement of the erosion of the complement of the image using the same SE [123]. However, the effect of dilation on a binary image is the opposite of that of erosion; dilation expands the objects in a binary image, whereas erosion shrinks the objects in a binary image. Similar to erosion, the dilation of a set, X, using an SE, B, can be represented in terms of the union of the two sets as

$$\delta_B(X) = \bigcup_{b \in B} X_{-b}, \tag{3.5}$$

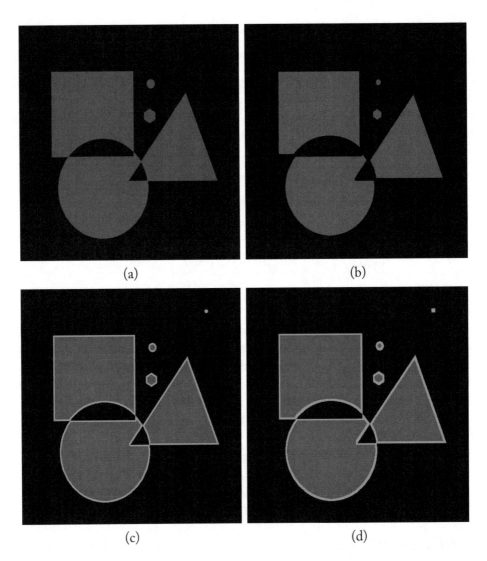

Figure 3.2: (a) A test image, of size 584×565 pixels, with several objects of different geometrical shapes. (b) The result of eroding the image in (a) using a disk-shaped SE with a radius of $r = 5$ pixels. (c) The same image as in (b) showing the regions that were eroded by the SE in green; the SE is also shown in green at the top-right corner of the image. (d) The result of erosion of the image in (a) using a square-shaped SE with side $s = 10$ pixels. The SE is shown in green at the top-right corner of the image; the eroded regions are also shown in green. Note that erosion shrinks the objects in an image; the amount of erosion is dependent on the shape and size of the SE.

where b determines the translation of B. The result of dilation can be considered as a set of points, x, where B will touch X, if the origin of B is placed at x. In mathematical terms,

$$\delta_B(X) = \{x \mid B_x \cap X \neq 0\}, \tag{3.6}$$

where B_x denotes the SE centered at x. For practical implementation, binary dilation can be formulated as

$$\delta_B[f(x)] = \max_{b \in B} f(x + b), \tag{3.7}$$

where f is the given image. Equation 3.7 implies that dilation changes the value of a given pixel x, to be equal to the maximum value of the image f in the area defined by the SE, when B is centered at x. In other words, the pixel where B is centered will be dilated (turned on) if at least one pixel from the binary image, f, is equal to one in the area defined by B. The binary dilation of a set, X, using an SE, B, is equivalent to the Minkowski addition of B to X if and only if the SE is symmetric ($B = \check{B}$) [123]. The derivation of the previous statement is similar to the derivation of binary erosion as Minkowski subtraction as explained previously.

Figure 3.3 shows the effect of dilation on the same test image as in Figure 3.2 (a) using different SEs. The square-shaped SE has a larger area than the disk-shaped SE; hence, dilation using the former causes more expansion of the objects in the image as compared to dilation using the latter. This point is evident by comparing the green areas in Figure 3.3 (d) to those in part (c) of the same figure.

In the present work, binary dilation is used to obtain border pixels of the effective area of the masks of the images of the DRIVE database in a preprocessing step, as explained in Section 5.2.

3.1.3   BINARY OPENING AND CLOSING

The morphological operation of erosion followed by dilation is called the opening operation [122, 123]. Binary opening is useful for the removal of small unconnected segments in a binary image or for pruning binary skeletons. As the name suggests, opening also can be used for opening small gaps between two connected segments of a pattern. Depending on the SE used, erosion eliminates unwanted parts that connect the two segments; dilation then restores parts of the pattern that were wanted, but were lost during the erosion operation. The morphological operation of closing is the dual operator of opening with respect to complementation [122, 123]. It consists of dilation of the image followed by erosion. Binary closing can be useful in filling small holes and segments in a binary image; dilation fills out incomplete parts of a pattern based on the shape of the SE, and erosion removes the parts that were dilated, but did not need to be.

Figure 3.4 shows the effects that the processes of binary opening and closing have on geometric shapes. A disk-shaped SE with radius $r = 7$ pixels is large enough to remove the four small circles and has opened a gap on the right-hand side between the circle and the square with binary opening, as shown in Figure 3.4 (b). While performing binary closing, as shown in part

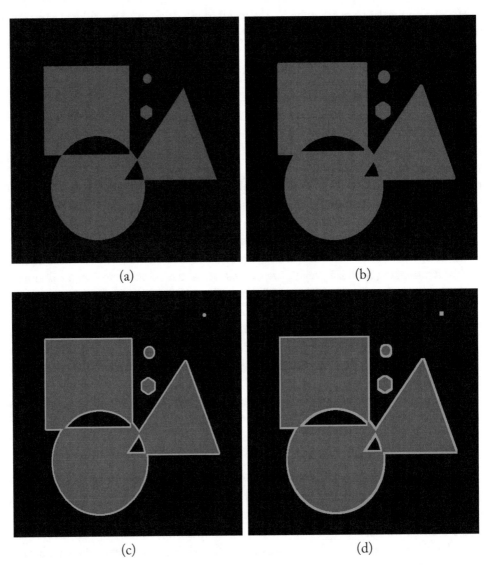

(a)

(b)

(c)

(d)

Figure 3.3: (a) A test image, of size 584×565 pixels, with several objects of different geometrical shapes. (b) The result of dilating the image in (a) using a disk-shaped SE with a radius of $r = 5$ pixels. (c) The same image as in (b) showing the regions that were dilated by the SE in green; the SE is shown in green at the top-right corner of the image. (d) The result of dilation of the image in (a) using a square-shaped SE with side $s = 10$ pixels. The SE is shown in green at the top-right corner of the image; the dilated regions are also shown in green. Note that dilation expands the objects in an image; the amount of expansion depends on the shape and size of the SE.

(c) of the same figure, the SE has filled the small circular- and pentagonal-shaped holes within the large square, but it has failed to fill the bigger rectangle-shaped hole. Both gaps between the triangle and the circle are also closed by the SE, when performing binary closing.

3.1.4   SKELETONIZATION

The term skeletonization refers to several different algorithms that extract a skeleton, S, which is a set of one-pixel-thick piecewise curvilinear center-lines of the patterns in the given binary image. Let P denote the white pixels of the pattern that have a value of 1; let B denote the black pixels of the background that have a value of 0. The main property of S, which is independent of the algorithm used, is that it is spatially situated along the medial region of P, that is, the distances from each pixel on S to at least two boundary pixels of P are equal [123, 125]. However, depending on the algorithm used to obtain the skeleton, the results may possess several other properties that suit different applications. Medial-axis transformation (MAT) [125] defines a skeleton as a set of pixels on P, where if a circle of a certain radius were to be centered, it would fit entirely within P. The MAT also provides the distances from each skeleton point to B as the radius of the previously defined circle. However, the result of the MAT may not always reflect the exact topological properties of the patterns in a given image.

A thinning algorithm produces a skeleton that is the union of simple arcs and curves and is independent of small inflections in the contours of P [126]. Hence, a thinning algorithm emphasizes the linear structure of patterns in an image.

The curvature-skeleton algorithm is similar to a thinning algorithm, except that it accounts for small protrusions and curvatures that might exist in the patterns [125]. The labeled-skeleton algorithm [125] provides the distances from each pixel on S to B, as well as the radius of the maximal discs, as in the MAT. The result of the labeled-skeleton algorithm is a more accurate representation of the topology of the pattern and provides accurate measurements of the thickness of the pattern as compared to the MAT.

A topology-preserving skeletonization algorithm guarantees that P does not fall apart in the process of skeletonization, that is, no holes are created in S. For this purpose, the m-connectedness of the pattern must not change. In order to explain a pixel-removal algorithm, several notions need to be described first. A pixel, p, has eight immediate neighbors, as shown in Figure 3.5. Assuming that p belongs to P (has a value of one), and given that $N(p)$ represents the summation of the values of the immediate neighboring pixels, p is said to be 8-adjacent to a neighbor pixel if $N(p) \geq 1$ for $1 \leq k \leq 8$ (p has at least one neighbor that belongs to P). The pixel p is 4-adjacent to a neighbor pixel if $N(p) \geq 1$ for $k = 1, 3, 5$, and 7. P is said to be 8- or 4-connected if it cannot be partitioned into two subsets that are not 8- or 4-connected, respectively. The choice of $m = 4$ is suitable for patterns that contain sharp corners and transitions, whereas $m = 8$ is appropriate for other general patterns [127]. An m-component is a subset of pixels of P that are m-connected and are not m-adjacent to any other pixels [127], as demonstrated in Figure 3.6. The contour, C, of a pattern P is a subset of pixels that are neighbors of

Figure 3.4: (a) A binary test image, of size 584×565 pixels, featuring various geometric shapes; the small circles can represent unwanted objects or holes. The effects of binary opening and closing using a disk of radius $r = 7$ pixels are shown in (b) and (c), respectively. It can be observed in the result of opening in (b) that the SE is large enough to remove the four small circles and has opened a gap between the circle and the square. In an opposite fashion, with reference to the result of closing in (c), the SE is large enough to fill the small circular- and pentagonal-shaped holes within the large square, but it has failed to fill the bigger rectangle-shaped hole. The gaps between the triangle and the circle are also closed in (c).

B. Iterative algorithms may be applied in parallel or sequential order to P to remove pixels from C in order to obtain a skeleton that does not fall apart. A pixel p, that belongs to C, is said to be m-deletable if its removal does not change the number of m-components that make up P and does not create a hole in P. It has been shown that if the removal of a pixel p does not change the topology of a 3×3 window centered on p, the global topology of P is also preserved [128]. There are various arithmetic operations that can determine if disconnectedness results, if a pixel, p, is removed, by considering a 3×3 neighborhood centered on p; the crossing [129, 130] and the connectivity numbers [131] are two such tests. The crossing and the connectivity numbers using 8-connectedness are defined as

$$X_8(p) = \sum_{k=1}^{4} h_k, \text{ where } h_k = \begin{cases} 1, & \text{if } n_{2k-1} = 0 \text{ and } (n_{2k} = 1 \text{ or } n_{2k+1} = 1) \\ 0, & \text{otherwise,} \end{cases} \tag{3.8}$$

and

$$C_8(p) = \sum_{k=1}^{4} (\bar{n}_{2k-1} - \bar{n}_{2k-1} \times \bar{n}_{2k} \times \bar{n}_{2k+1}), \tag{3.9}$$

respectively, where $\bar{n}$ is the complement of n (that is, $\bar{n} = 1 - n$) and $n_9 = n_1$. Equations 3.8 and 3.9 both count the number of 8-components of white pixels in $N(p)$ in different manners. A pattern P will not be disconnected by the removal of p, if p is an 8-neighbor of only one 8-component in $N(p)$. A crude, yet simple skeletonization algorithm, using the 8-connectivity number (Equation 3.9), works as follows: remove a pixel p, that belongs to C, only if $X_8 == 1$; repeat until P does not change, that is, when P and C are equal.

n_2	n_3	n_4
n_1	p	n_5
n_8	n_7	n_6

Figure 3.5: Illustration of the adjacent neighborhood of a given pixel, p, in an image. $N(p) = \{n_k \mid 1 \le k \le 8\}$ determines 8-adjacency and $N(p) = \{n_k \mid k = 1, 3, 5, 7\}$ decides 4-adjacency, where n_k is a neighboring pixel value, with the value of one for a pixel that belongs to P and zero for a pixel that resides in B.

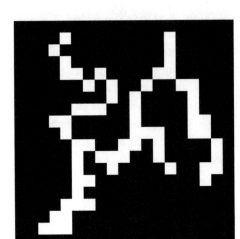

Figure 3.6: An image (19 × 19 pixels) demonstrating a pattern (white pixels) that is 8-connected as a whole and is made up of fifteen 4-component segments. Each 4-component segment is either an isolated pixel or a 4-connected set of pixels, but no 4-component segment has a pixel that is 4-adjacent to the pixel(s) of other 4-component segments.

The level of perceptual detail that is preserved in the process of skeletonization depends on accurate detection and identification of the different elements of a skeleton. A skeleton pixel can be a normal point, an end point, or a branch point. End points are extremes of arcs and curves (tips of pattern protrusions); branch points are common points of two or more arcs/curves. Normal points are pixels that are neither branch points nor end points. A thinning algorithm may not detect all protrusions and branch points [126], whereas a curvature-skeleton algorithm could detect all dominant points related to branch and end points [125]. There are several other more sophisticated arithmetic conditions that also can check for the connectedness of a pattern, as well as check if a pixel p is a normal point, a branch point, or an end point [125, 126, 132, 133].

Iterative algorithms, as explained above, could be inefficient for a fast implementation. A lookup table (LUT) could be used to indicate whether a pixel p should be deleted or not, based on the pattern of pixels in a 3 × 3 window centered on p and the type of skeletonization algorithm being used. Such an implementation is significantly faster than an iterative algorithm. In the present work, MATLAB's implementation of the curvature-skeleton algorithm, which uses LUTs to perform the skeletonization procedure, is used [134].

Figures 3.7 (b) and (c) show the difference between the results of the curvature-skeleton and thinning algorithms, respectively, as applied to the test image in part (a) of the same figure. It should be noted that not all the skeletons of geometric shapes convey intuitive visual information

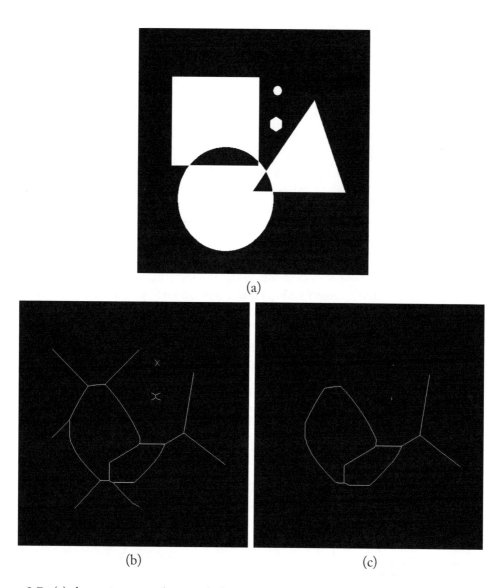

Figure 3.7: (a) A test image with several objects of different geometrical shapes. (b) The result of skeletonization of the image in (a) using the curvature-skeleton algorithm. (c) The result of skeletonization of the image in (a) using the thinning algorithm; it is evident that the result of the thinning algorithm does not preserve all of the topological details of the original patterns.

about the patterns under analysis; skeletons are particularly useful in the analysis of oblong and branching patterns.

In the present work, the curvature-skeleton algorithm is used to obtain the VST of the MTA by considering 8-connectivity. The details of application of the preprocessing step that uses the skeletonization procedure are provided in Section 6.1.

3.1.5   AREA OPEN OPERATOR

The morphological operation of area open locates m-component sets of white pixels in a binary image and removes those sets that are smaller than a specified area, which is given in terms of a certain number of pixels [135]. The area-close operation works in the same manner as the area-open operation, but removes sets of black pixels instead of white pixels.

Figure 3.8 shows the result of applying the area-open procedure to the 8-connected pattern shown in Figure 3.6 by considering 4-connectivity. It can be seen in Figure 3.8 (b) that all 4-component segments with area less than four pixels are removed.

Figure 3.9 shows the effect of applying the area-open operation by considering 8-connectivity to the test image with different types of geometrical shapes. Figure 3.9 (b) shows that the only subset of white pixels that has an area less than 377 pixels is the small circle, which is removed from the original image in part (a) of the same figure.

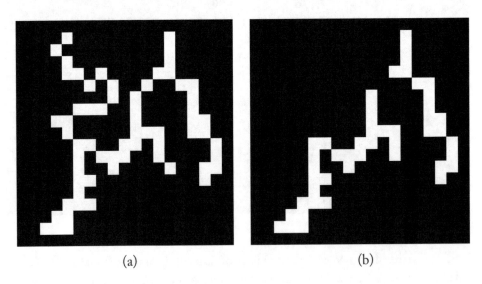

(a)   (b)

Figure 3.8: (a) An image (19×19 pixels) demonstrating a pattern (white pixels) that is made up of fifteen 4-component segments. (b) The result of applying the area-open operation to the image in (a); 4-connectivity is used to find 4-component segments in the pattern that have an area of less than four pixels and remove them.

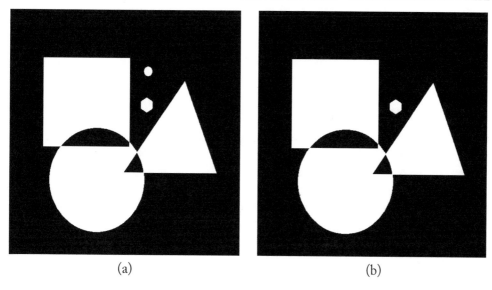

(a)   (b)

Figure 3.9: (a) A test image with several objects of different geometrical shapes. (b) The result of applying the area-open procedure on the image in (a); 8-connectivity was used to find subsets of white pixels that have an area less than 377 pixels. Only the small circle is removed from the original image.

In the present work, the morphological operation of area open is used in two instances to remove unwanted segments of white pixels resulting from the binarization process, as explained in Sections 5.2 and 6.1.

3.2   DETECTION OF ORIENTED PATTERNS

3.2.1   MATCHED FILTERS

Given a priori knowledge of the spatial properties of blood vessels, matched filters can serve as efficient tools for the extraction of blood vessels in retinal images [71, 76, 77, 136, 137]. The main concept behind the design of matched filters for the detection of blood vessels is that cross-sections of the intensity profile of the blood vessels are known to have a Gaussian shape. Based on this observation and other physiological and spatial properties of the blood vessels in retinal images, templates (filters) can be designed to match the intensity profile of the blood vessels. Some of the important design aspects of matched filters for the detection of blood vessels are the following:

1. Blood vessels have lower reflectance than the retinal surface; hence, they are darker than the background and have a negative contrast.

2. Blood vessel width decreases when moving away from the ONH and hence, the STD of the Gaussian models could be decreased to match the vessel width.

3. The average width of the retinal blood vessels is known, which helps to select a specific range for the STD of the Gaussian models.

The result of matched filtering of a retinal image is a grayscale image, which may be thresholded to produce a binary image of the detected blood vessels.

3.2.2   GABOR FILTERS

Gabor filters are sinusoidally modulated Gaussian functions that have optimal joint resolution in both the frequency and space domains [138, 139]. There has been a significant amount of research conducted on the use of Gabor functions for texture segmentation, analysis, and discrimination [140]. Gabor functions have also been found to provide good models for the receptive fields of simple cells in the striate cortex [141–143].

Gabor filters are essentially directionally selective bandpass filters, and may be used as line detectors [144, 145]. Gabor filters also have been shown to be efficient tools for the detection of blood vessels [85, 86, 146–148]. Gabor filters are divided into two categories: the real and the complex Gabor filters. The real Gabor filters have a higher optimal-scale range and are less sensitive to the presence of Gaussian noise, as compared to complex Gabor filters and other oriented feature detectors [144]; the real Gabor filters are also computationally less expensive than the complex Gabor filters.

The impulse response of the main real Gabor filter kernel oriented at the angle $\theta = -\pi/2$ may be formulated as [144, 149]

$$g(x, y) = \frac{1}{2\pi\sigma_x\sigma_y} \exp\left[-\frac{1}{2}\left(\frac{x^2}{\sigma_x^2} + \frac{y^2}{\sigma_y^2}\right)\right] \cos(2\pi f_o x), \tag{3.10}$$

where σ_x and σ_y are the STD values in the x and y directions, respectively, and f_o is the frequency of the modulating sinusoid.

The frequency response of the Gabor filter kernel given in Equation 3.10 is

$$G(u, v) = \frac{1}{2}\left(\exp\left[-2\pi^2\left(\sigma_x^2(u + f_o)^2 + \sigma_y^2 v^2\right)\right] + \exp\left[-2\pi^2\left(\sigma_x^2(u - f_o)^2 + \sigma_y^2 v^2\right)\right]\right). \tag{3.11}$$

A basis set can be obtained through rotation of the main kernel using the rotation matrix given by [150]

$$\begin{bmatrix} x' \\ y' \end{bmatrix} = \begin{bmatrix} \cos(\theta) & -\sin(\theta) \\ \sin(\theta) & \cos(\theta) \end{bmatrix} \begin{bmatrix} x \\ y \end{bmatrix}. \tag{3.12}$$

The parameters in Equation 3.10, namely σ_x, σ_y, and f_o, are derived by taking into account the size of the lines or curvilinear structures to be detected. Let τ be the thickness of the line detector. This parameter is related to σ_x and f_o as follows [86]: the amplitude of the exponential

(Gaussian) term in Equation 3.10 is reduced to one half of its maximum at $x = \tau/2$ and $y = 0$; therefore, $\sigma_x = \tau/(2\sqrt{2\ln 2}) = \tau/2.35$. The cosine term has a period of τ; hence, $f_o = 1/\tau$. The value of σ_y is defined as $\sigma_y = l\,\sigma_x$, where l determines the elongation of the Gabor filter in the direction of the defined orientation, with respect to its thickness. The values of τ and l could be varied to prepare a bank of filters at different scales for multiscale filtering and analysis. The specification of the Gabor filter in terms of the parameters $\{\tau, l\}$ facilitates easier design and analysis in relation to the size of the features to be detected than the use of the parameters $\{\sigma_x, \sigma_y, f_o\}$. The real Gabor filter designed as above with a cosine term can detect oriented features of positive contrast, that is, elements that are brighter than their immediate background. The complex Gabor filter could be used to detect features of unknown contrast because it has both a sine and a cosine term.

Let $I(x, y)$ be the image being processed and $\phi(x, y)$ be the angle of the blood vessels at (x, y). Let $g_k(x, y), k = 0, 1, ..., (K-1)$ form a bank of real Gabor filters oriented at $\alpha_k = -\pi/2 + \pi k/K$ angles, where K is the number of equally separated filters over the range $[-\pi/2, \pi/2]$, for a given pair of values of $\{\tau, l\}$. The Gabor-filtered image is given by $W_k(x, y) = (I * g_k)(x, y)$, where the asterisk denotes linear convolution. The orientation field of $I(x, y)$, produced by the bank of real Gabor filters, is given by both the angle $\phi(x, y) = \alpha_{k_{max}} + \frac{\pi}{2}$, where $k_{max} = \arg\{\max[W_k(x, y)]\}$, and by the magnitude of the output of the real Gabor filter at the optimal orientation, $M(x, y) = W_{k_{max}}(x, y)$ [144].

Figure 3.10 shows Gabor filters for various values of the parameters τ, l, and θ, demonstrating the effects of scaling, stretching, and rotation; the corresponding frequency-domain magnitude transfer functions are shown in Figure 3.11. It is evident that a Gabor filter acts as a bandpass filter, with a limited range of response that is dependent upon the parameters and orientation of the filter. Decreasing the scale factor τ causes the filter to shift to a higher frequency band. Reducing the elongation factor l causes the filter to be less directionally sensitive. Rotating a filter causes a corresponding rotation of the frequency response. Note that there is a $\pi/2$ shift between angles in the space domain (Figure 3.10) and the frequency domain (Figure 3.11).

Gabor filters are used in the present work for the detection of blood vessels in retinal images, as described in Chapter 5.

3.2.3   COHERENCE

Rao and Schunck [151] characterized the information related to oriented features in images in terms of intrinsic orientation angle and coherence images. They used a gradient-of-Gaussian filter to obtain the gradient magnitude and the gradient orientation at a pixel (m, n) in an image as

$$G_{mn} = \sqrt{G_x^2(m, n) + G_y^2(m, n)}, \qquad (3.13)$$

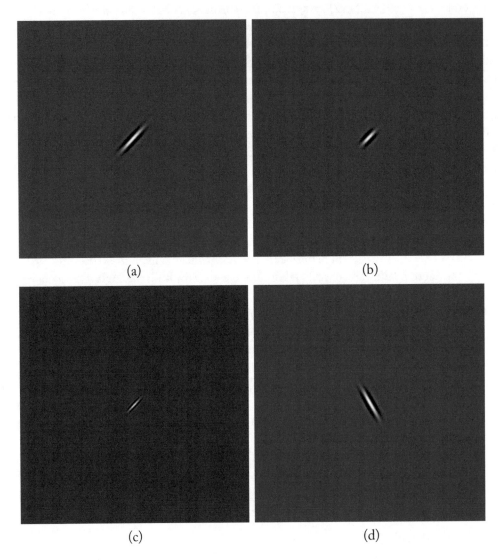

Figure 3.10: Gabor filters for various values of the parameters τ, l, and θ, demonstrating the effects of scaling, stretching, and rotation. (a) $\tau = 8$, $l = 2.9$, $\theta = 45°$. (b) $\tau = 8$, $l = 1.7$, $\theta = 45°$. (c) $\tau = 4$, $l = 2.9$, $\theta = 45°$. (d) $\tau = 8$, $l = 2.9$, $\theta = -60°$. Each filter was created using a matrix of size 256×256 pixels. τ is given in pixels, unless otherwise specified. See Figure 3.11 for the corresponding frequency-domain magnitude responses.

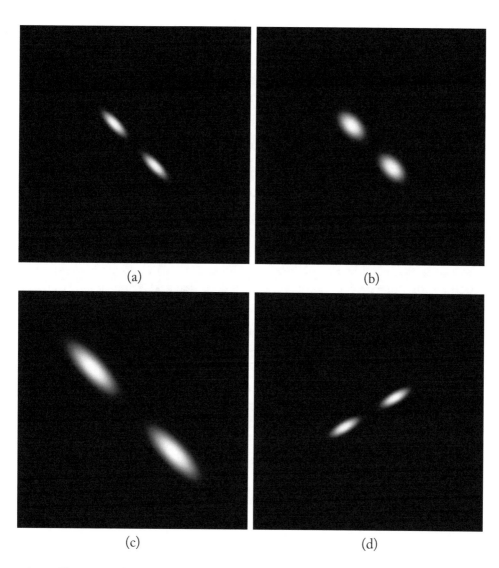

(a)

(b)

(c)

(d)

Figure 3.11: Frequency-domain magnitude transfer functions of the Gabor filters shown in Figure 3.10 for various values of the parameters τ, l, and θ, demonstrating the effects of scaling, stretching, and rotation. Each function was created using a matrix of size 256×256 pixels. The DC or $(0,0)$ frequency component is at the center of the frequency response image in each case. (a) $\tau = 8, l = 2.9, \theta = 45°$. (b) $\tau = 8, l = 1.7, \theta = 45°$. (c) $\tau = 4, l = 2.9, \theta = 45°$. (d) $\tau = 8, l = 2.9, \theta = -60°$.

and

$$\theta_{mn} = \arctan\left(\frac{G_y(m,n)}{G_x(m,n)}\right), \tag{3.14}$$

respectively, where G_x and G_y represent the outputs of the gradient-of-Gaussian filter in the x and y directions, respectively. Rao and Schunck [151] estimated the oriented angle of flow in a $P \times P$ neighborhood centered at a pixel located at (p, q) as

$$\psi_{pq} = \frac{1}{2}\arctan\left(\frac{\sum_{m=1}^{P}\sum_{n=1}^{P} G_{mn}^2 \sin(2\theta_{mn})}{\sum_{m=1}^{P}\sum_{n=1}^{P} G_{mn}^2 \cos(2\theta_{mn})}\right) + \frac{\pi}{2}, \tag{3.15}$$

because the gradient angle is perpendicular to the direction of flow. They computed the coherence at the pixel (p, q) in the given image, using a window of size $P \times P$ as the cumulative sum of the projections of the gradient magnitudes in the direction of the dominant flow at (p, q), which was normalized by the cumulative sum of the gradient magnitude values in the specified window. The result was then multiplied by the gradient magnitude at (p, q) to obtain high coherence values for patterns with a preferred or dominant orientation. Rao and Schunck [151] formulated the coherence γ_{pq} at point (p, q) as

$$\gamma_{pq} = G_{pq}\frac{\sum_{m=1}^{P}\sum_{n=1}^{P} G_{mn} \cos(\theta_{mn} - \psi_{pq})}{\sum_{m=1}^{P}\sum_{n=1}^{P} G_{mn}}. \tag{3.16}$$

The use of a gradient filter, such as the gradient-of-Gaussian filter, is appropriate if the edge information is adequate to characterize the oriented features in a given image. In the present work, the magnitude and orientation information regarding the blood vessels in a given retinal fundus image is obtained using Gabor filters, as described in Section 3.2.2. Thus, the estimated angle of flow and the coherence images are obtained as

$$\psi_{pq} = \frac{1}{2}\arctan\left(\frac{\sum_{m=1}^{P}\sum_{n=1}^{P} M_{mn}^2 \sin(2\phi_{mn})}{\sum_{m=1}^{P}\sum_{n=1}^{P} M_{mn}^2 \cos(2\phi_{mn})}\right), \tag{3.17}$$

and

$$\gamma_{pq} = M_{pq}\frac{\sum_{m=1}^{P}\sum_{n=1}^{P} M_{mn} \cos(\phi_{mn} - \psi_{pq})}{\sum_{m=1}^{P}\sum_{n=1}^{P} M_{mn}}, \tag{3.18}$$

respectively, where M and ϕ represent the Gabor magnitude and Gabor angle outputs, respectively, as described in Section 3.2.2.

In the present work, coherence is used as a feature to enhance the results of the detection of blood vessels by performing multifeature analysis using pattern classification methods, as described in Section 5.5.

3.2.4   ILLUSTRATIONS OF DETECTION OF ORIENTED PATTERNS

Figure 3.12 illustrates the effects of applying Gabor filters with different design parameters to a test image of tree branches. Figure 3.12 (a) shows a part of the original color image. Branching patterns are known to be piecewise-linear oriented features. The real Gabor filter, as used in the present work, requires a grayscale image where the oriented features have a positive contrast. Hence, the inverted luminance component of the YIQ color space model (see Section 5.2) is used, as shown in Figure 3.12 (b). For the results in Figures 3.12 (c), (d), and (e), the orientation of the Gabor function was kept constant at 45°; only branches that are oriented at 45° have been detected with prominent response. Figure 3.12 (d) shows the effect of decreasing the thickness parameter, τ, as compared to part (c) of the same figure. Similarly, Figure 3.12 (e) shows the effect of decreasing the elongation parameter, l, as compared to part (c). It is evident that different values of τ and l detect branches with different thickness and elongation. Figure 3.12 (f) shows the effect of changing the orientation of the Gabor filter, as compared to part (c) of the same figure; only branches that are oriented at 10° have been detected.

To obtain the results shown in Figures 3.13 (a), (b), and (c), a bank of 180 Gabor filters was used over the range of $[-\pi/2,\ \pi/2]$ (one filter per step of 1°) to obtain the Gabor magnitude responses. Because the thickness of the branches is not always the same, different values of τ have detected branches of different thickness; $\tau = 4$ has led to the detection of smaller and thinner branches with high response, $\tau = 12$ has resulted in the thick branches of the tree being emphasized, whereas $\tau = 8$ has resulted in the detection of most of the branches, but not the thinnest or the thickest branches. These observations imply the need for multiscale analysis using different scales of thickness for accurate detection of all of the branches in the test image. Figure 3.13 (d) shows the angles related to part (c) of the same figure, shown in the form of needles for every fifth pixel, for a part of the image. The Gabor magnitude and angle responses obtained using the Gabor parameters in Figure 3.13 (b) were used to acquire the coherence images in Figures 3.13 (e) and (f); the choice of the $P \times P$ neighborhood used to compute the average angle of flow at each point in the image has an effect on the resulting coherence images.

Figures 3.14 (a) and (b) show the effect of binarization of the image in Figure 3.13 (b) using a threshold obtained automatically using Otsu's optimal thresholding method [152] and a manually set threshold, respectively. Any pixel in the grayscale image with an intensity value higher than the specified threshold is set to one in the binarized image; any pixel with an intensity value lower than the specified threshold is set to zero in the resulting binary image. The choice of an optimal threshold is crucial to all subsequent steps, as it can introduce unwanted artifacts if the threshold is set too high or omit parts of the desired pattern if it is set too low, as shown in Figure 3.14 (b). Part (c) of Figure 3.14 shows the effect of applying binary erosion, using a disk-shaped SE of radius one pixel, to the image shown in (a). Note that three-pixel-wide branches are eroded to a thickness of one pixel; thinner segments are entirely eroded. Figure 3.14 (d) shows the result of performing binary dilation, using a disk-shaped SE of radius one pixel, on the oriented branching pattern in part (a) of the same figure. Note that the dilation process has made

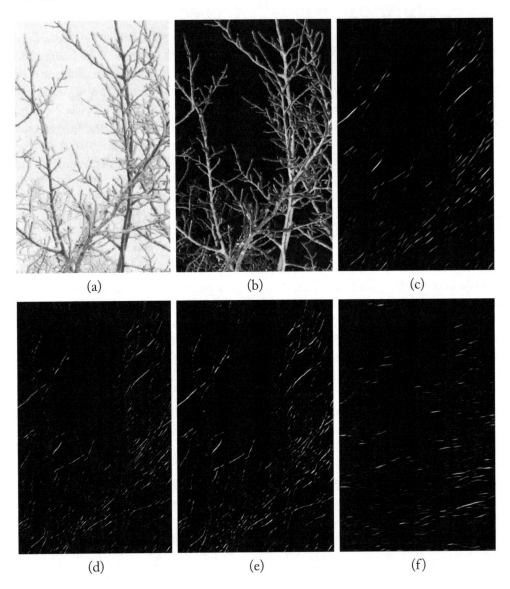

Figure 3.12: (a) A test image, of size 764×486 pixels, showing tree branches, which are considered to be piecewise-linear features. (b) The inverted Y component of the YIQ color space (see Section 5.2) is used as the input to the Gabor filters so that the oriented features have a positive contrast in the input image. Gabor magnitude responses for the image in (b) for different values of τ, l, and θ are illustrated as follows: (c) $\tau = 8$, $l = 1.8$, $\theta = 45°$; (d) $\tau = 4$, $l = 1.8$, $\theta = 45°$; (e) $\tau = 8$, $l = 0.9$, $\theta = 45°$; and (f) $\tau = 8$, $l = 1.8$, $\theta = 10°$.

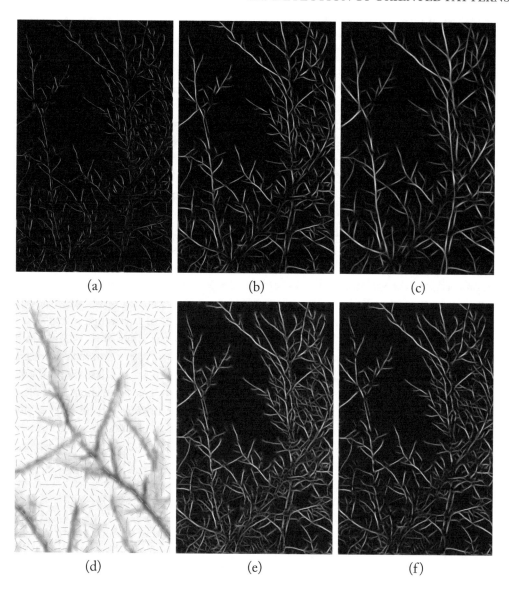

(a)   (b)   (c)

(d)   (e)   (f)

Figure 3.13: Gabor magnitude responses for the image in Figure 3.12 (b) over the range $[-\pi/2, \pi/2]$ in steps of $1°$ (180 Gabor filters) are shown for different values of τ, with $l = 1.8$: (a) $\tau = 4$, (b) $\tau = 8$, and (c) $\tau = 12$. (d) Gabor angles related to (b), shown in the form of needles for every fifth pixel for a small part of the image. Coherence images, as obtained using the Gabor magnitude and angle images in (b) using disk-shaped neighborhoods of (e) radius $r = 3$ pixels, and (f) radius $r = 4$ pixels to compute the average angle of flow at each pixel.

all segments in Figure 3.13 (a) two pixels thicker, as shown in part (d) of the same figure. Figures 3.14 (e) and (f) show the effects of applying the opening and closing operations. Note that the opening operation [Figure 3.14 (e)] removes small segments that are present in part (a) of the same figure, and also breaks apart narrow branches. On the other hand, the closing operation [Figure 3.14 (f)] fills small holes that exist in part (a) of the same figure; however, it also fills small empty segments that are enclosed between adjacent branches.

Figure 3.15 (b) shows the effect of applying the curvature-skeleton algorithm to the test image with oriented branching patterns, shown in part (a) of the same figure. The resulting skeleton is oriented in the same direction as the branching patterns and lies in the medial regions of the branches. Part (c) of the same figure shows the result of applying the thinning algorithm to the image in (a); the skeletons in (b) and (c) may appear to be similar at first, but on close inspection, it can be observed that the skeleton obtained using the thinning algorithm does not have forked end points at the tips of the branches, as does the skeleton obtained using the curvature-skeleton method. Also, curved branches in the skeleton in (c) are constructed of more linear segments, as compared to the skeleton in (b). Figure 3.15 (d) demonstrates the effects of applying the area-open procedure to the skeletonized image of branching patterns, as shown in part (b) of the same figure; 8-connected segments of white pixels with an area less than 20 pixels are removed in (d). It can be seen that the small artifacts on the left-hand border of the image in part (b) are not present in the result in part (d). Figures 3.15 (e) and (f) show the effect of varying the minimum number of connected pixels required for deletion, as well as the type of connectivity used by the area-open procedure, respectively. Segments of white pixels with an area less than 70 pixels that are 8-connected are removed in (e); most of the artifacts, as well as some short branches are removed by this process. Figure 3.15 (f) shows the effect of removing 4-connected segments of white pixels with an area less than 10 pixels; the branching pattern is reduced to small horizontal and vertical segments, as one-pixel-thick diagonally connected patterns are not considered to be connected with 4-connectedness.

3.3   DETECTION OF GEOMETRICAL PATTERNS

The HT has long been recognized as a powerful image processing method for the detection of curves, shapes, and motion in images with noisy, irrelevant, or even missing data [153, 154]. Hough [155] originally proposed a method for the detection of straight lines in bubble-chamber photographs. The method has since been modified and extended in many different ways to detect lines, circles, and parabolic and hyperbolic curves; for estimation of 2D and three-dimensional (3D) motion; for object recognition; and for detection of arbitrary shapes [153, 154, 156]. The HT either refers to the general process of detection of shapes, or to the original method to detect straight lines and its different variations; the GHT refers to methods that employ the basic principles of the HT process, but detect other shapes and curves. Both the HT and the GHT have been used in industrial settings, as well as in image processing hardware algorithms for rapid detection of lines and other curves [153]. One of the earliest applications of the GHT in bio-

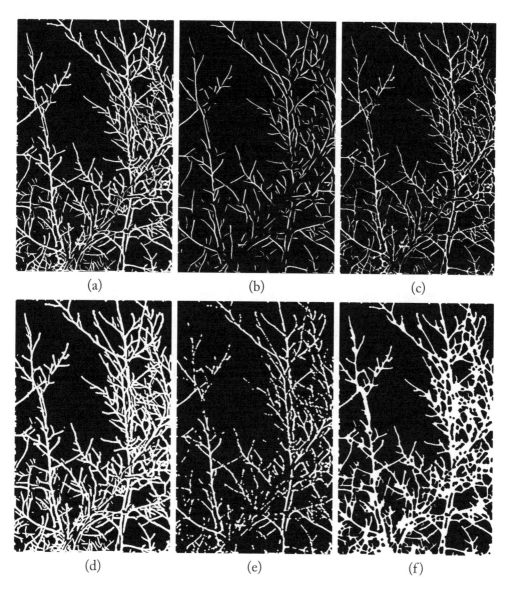

(a)   (b)   (c)

(d)   (e)   (f)

Figure 3.14: The result of binarization of the image in Figure 3.13 (b), using (a) Otsu's optimal thresholding method and (b) manual thresholding. (c) The result of applying binary erosion using a disk of radius $r = 1$ pixel to the image in (a). (d) The result of dilating the image in (a) using the same SE as in (c). (e) Result of applying binary opening using a disk of radius $r = 2$ pixels to the image in (a). (f) Result of applying the binary closing operator to the image in (a) using a disk of radius $r = 4$ pixels.

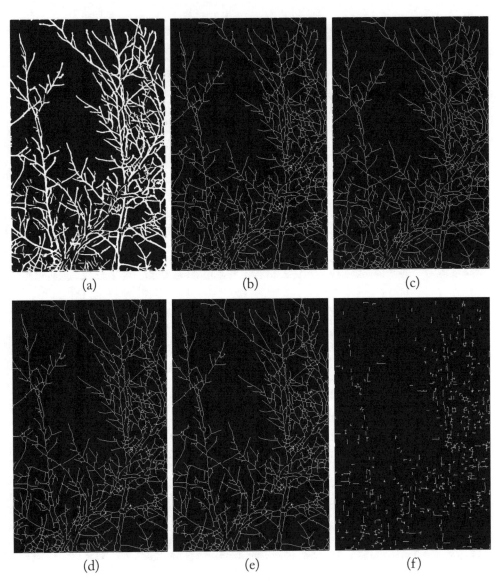

Figure 3.15: (a) The same image as in Figure 3.14 (a). (b) The skeleton of the image in (a), obtained using the curvature-skeleton algorithm. (c) The skeleton of the image in (a), obtained using the thinning algorithm. The result of applying the area-open procedure to the image in (b) to remove connected segments of white pixels using (d) 8-connectedness to remove segments that have an area less than 20 pixels, (e) 8-connectedness to remove segments that have an area less than 70 pixels, and (f) 4-connectedness to remove parts that have an area less than 10 pixels.

medical image processing was demonstrated by Wechsler and Sklansky [157]; they applied the GHT for the detection of parabolas to X-ray images of the chest to detect the rib cage. Different forms and variations of the GHT have since been used in various biomedical image processing applications [108, 132, 158–162].

In general, the HT has several important properties that make it desirable for shape detection, as follows:

1. The HT recognizes partial or slightly misshaped curves and lines, which can be a difficult task for other shape detection methods.

2. The HT detects the specific shape in the presence of random and unrelated data.

3. The HT detects several variations of the same shape in the image in one operation.

4. The HT processes each pixel independently; hence, parallel processing of data is possible for fast hardware implementation.

It should be noted that unwanted data that are similar to the shape to be detected could have an effect on the outcome of the HT, and should be treated with care.

The HT reduces a global shape detection problem in the spatial domain to a simpler global peak detection problem in a parameter space. Every spatial point that belongs to the pattern leads to a vote on different combinations of parameters that could have caused its presence, if it were part of the shape to be detected. An accumulator matrix is used to store and count the votes. The final count for each accumulator cell indicates the likelihood of the shape, described by the corresponding parameter values of the accumulator cell, belonging to the given pattern in the spatial domain. The size of the accumulator matrix is determined by the number of parameters and their range of values.

The HT is said to be similar to template-matching methods, but it is more efficient and advantageous [153, 154, 163]. Template matching is implemented entirely in the spatial domain, whereas the HT is implemented in the parameter space or the Hough space. In template matching, different templates are generated by shifting and reflecting a basic template and then trying to determine how well the image points match the template points. However, in many instances, corresponding points do not exist in the image domain, which makes the template-matching algorithm inefficient. The HT assumes a match between a given template and an image point and then attempts to determine the transformation parameters that relate the two. In other words, the HT does not generate the inessential data that are generated by a template-matching algorithm. Further restrictions on the limits of the Hough-space parameters can make the HT even more efficient.

Specific forms of the HT also have been shown to have similarities to other detection methods, such as the Radon transform [132, 164, 165] and the generalized maximum-likelihood estimators [153, 154]. The HT for line-segment detection has some qualitative properties that are similar to the Radon transform. However, the Radon transform cannot provide all the different variations that are possible to achieve using the HT. Depending on the kernel function used to relate the image in the spatial domain to the parameter space, the HT can take on different forms of the class of maximum-likelihood estimators. A quadratic kernel function, for example, implies that the HT could behave like a least-squares estimator in the continuous domain; the concept can be extended to the discrete domain [154].

The following sections provide details on the implementation of a simple form of the HT to detect straight lines, as well as details of applying a form of the GHT for the detection of parabolas; the latter is used to model the MTA in the present work.

3.3.1 THE HOUGH TRANSFORM FOR THE DETECTION OF LINES

Hough proposed a method for the detection of straight lines in bubble-chamber images based on the slope-intercept formulation [155]; this method was later refined by Duda and Hart [166], because of the limitations that the slope-intercept formula imposed. Duda and Hart used the normal parameters (θ, ρ) to represent a straight line as

$$\rho = x \cos(\theta) + y \sin(\theta), \tag{3.19}$$

where ρ is the length of the normal to the line from the origin and θ is its angle with respect to the horizontal axis. For the sake of simplicity, let the origin be at the center of the image. In such a representation, the parameter ρ is considered to be negative if the normal to the line extends below the horizontal axis ($y = 0$). (Note that this does not mean that the normal has a negative length.) The variable θ is the angle of the normal as measured from the horizontal axis line to the normal in the counterclockwise direction. Ideally, one-pixel-thick lines should be present in the input image to the HT, as multiple lines may exist within a thick line. The HT states that, for every pixel on the line in the image domain, there exists a sinusoidal curve in the Hough space. All sinusoidal curves that correspond to the pixels on the same line intersect at the same point, (θ_0, ρ_0), in the Hough space. Similarly, every point in the Hough space represents a line in the image domain. The size of the Hough space is determined by the limits of the (θ, ρ) parameters: the limits of ρ are determined by the size of the image as $L = \pm \frac{\sqrt{n^2 + m^2}}{2}$, where (m, n) are the height and width of the image, respectively; the parameter θ lies in the range $K = [0, \pi]$. It should be noted that it is possible to change the limits of the parameter ρ to be within $L = [0, \sqrt{n^2 + m^2}]$, in which case the limits of the parameter θ will change to be in the range $K = [0, 2\pi]$; in such a setup, negative values of ρ are not used. For every nonzero pixel in the given image, θ is varied within the specified range and Equation 3.19 is solved for the parameter ρ; if ρ falls within the specified range of L, then the corresponding accumulator cell in the Hough space, $[\theta(k), \rho(l)]$, is incremented by 1. The point in the Hough space that is incremented the most (has the highest value) is the point of intersection of the sinusoids that correspond to pixels on the same line, and provides the parameters of the detected line. Figure 3.16 illustrates an example of the HT as applied to a test image with two lines. Note that, in this case, there are two major points of intersection of the sinusoids in the Hough space [Figure 3.16(c)], which represent the two lines in

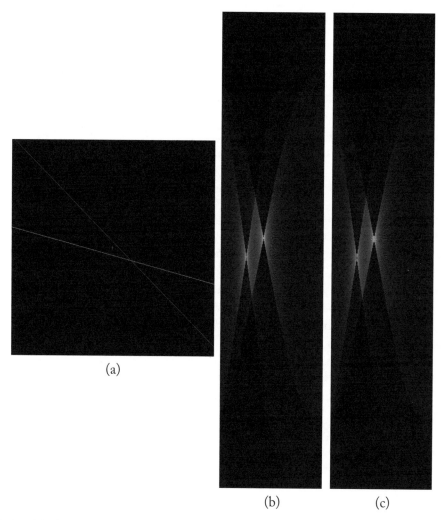

(a)

(b)   (c)

Figure 3.16: (a) A test image of size 584×565 pixels with two straight lines; the red line has $(\theta, \rho) = (45, 12)$ and the green one has $(\theta, \rho) = (75, -20)$. The limits of the horizontal and vertical axes are ± 283 and ± 292, respectively, with the origin at the center of image as annotated by the magenta '+' mark. (b) The Hough space for the image in (a); the two points with the highest and second-highest intensity values represent the parameters of the detected lines. The display intensity is $\log_{10}(1 + $ accumulator cell value). The limits of the horizontal and vertical axes are $\theta = [0, \pi]$ and $\rho = [-407, 407]$, with the origin at the center of the image. (c) The points $(\theta_o, \rho_o) = (45, 12)$, and $(75, -20)$, marked by the red and the green '*' symbols, represent the detected parameters of the red and the green line, respectively.

the spatial domain. The point representing the red line in the Hough space domain has a higher intensity than the point representing the green line, because the red line is made up of more pixels than the green line.

3.3.2   THE GHT FOR THE DETECTION OF PARABOLAS

The GHT can be used to detect parametric curves such as circles and parabolas [118, 132, 153, 156, 157, 163]. In order to implement the GHT, first, one must identify the parameters that govern the shape of the feature to be detected. The general formula defining a parabola with its directrix parallel to the vertical axis m and its symmetrical axis parallel to the horizontal axis n is

$$(m - m_o)^2 = 4a(n - n_o), \tag{3.20}$$

where (m_o, n_o) is the vertex of the parabola and the quantity $4a$ is known as the latus rectum. The absolute value of the parameter a governs the aperture or openness of the parabola, and its sign indicates the direction of the opening of the parabola; for a positive a value, the parabola opens to the right—see Figure 3.17 (a). The parameters (m_o, n_o, a) define the Hough space, represented by an accumulator, A. For every nonzero pixel in the image domain (m, n) [see Figure 3.17 (a)], there exists a parabola in the Hough space for each value of a that opens in the direction opposite to that of the parabola in the image domain [see Figures 3.17 (b)-(d)]; a single point in the Hough space defines a parabola in the image domain. The size of each (m_o, n_o) plane in the Hough space is defined to be the same as the size of the given image. The size of the parameter a is theoretically unbounded; however, in many applications, a limit can be imposed on the available or relevant range of the parameter a, based on physical limitations and/or by empirically determining a desirable range.

In the procedure implemented in the present work [118], for each nonzero pixel in the image shown in Figure 3.17 (b), the parameter a is computed for each (m_o, n_o) in the Hough space, and the corresponding accumulator cell is incremented by 1, if the value of a is within a specified range. The two points in the Hough space with the highest and the second-highest values are obtained as the parameters of the best-fitting parabolic models (for the example in Figure 3.17). The coordinates of each detected vertex, along with the corresponding a value, define a parabolic model. The observation of oppositeness of the direction of the parabolas between the image domain and the Hough space led to the generalization of the HT by Sklansky [163] as a form of an efficient template-matching algorithm.

In the present work, the GHT for the detection of parabolas is used to model the MTA, as described in Chapter 6.

3.4   PATTERN CLASSIFICATION

Machine learning algorithms can be used to distinguish different patterns in a given set of input values; this process is known as pattern recognition [167]. Pattern classification refers to a similar

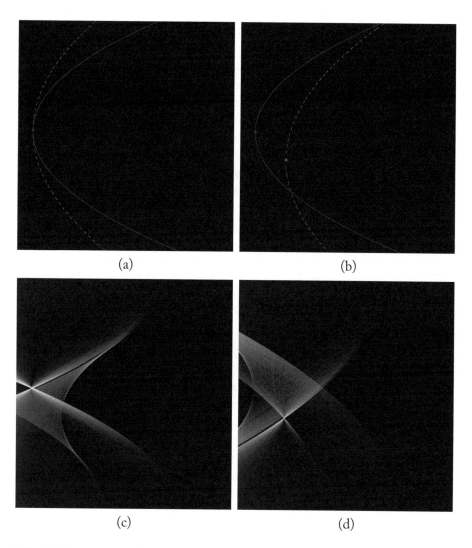

Figure 3.17: (a) A test image of size 584 × 565 pixels, showing two parabolas with different openness parameters of $a = 60$ (solid red line) and $a = 120$ (dashed green line). Both parabolas share the same vertex for the sake of comparison of their openness. (b) The same parabolas as in (a), but with different vertices; the solid red parabola has its vertex at (280, 40) and the dashed green parabola has its vertex at (350, 120); the top-left corner of the image is represented by the coordinates (1, 1). (c) The resulting Hough-space plane for the image in (b) for $a = 60$, showing the detected vertex, (280, 41), for the parabola with the smaller aperture as the point with the highest value. (d) The Hough-space plane for $a = 120$, showing the detected vertex, (351, 120), for the parabola with the larger aperture.

process that classifies a given set of inputs into a given set of classes. Binary pattern classification deals with grouping a given set of input values into exactly two classes; binary classifiers are widely used in biomedical applications, where it is desired to classify the data into normal and abnormal cases, that is, negative and positive diagnosis. Pattern classification methods may be divided into two main categories: linear and nonlinear classifiers. Linear classifiers, such as linear discriminant functions and single-layer perceptrons, work well if the available data are linearly separable into different classes. However, in many applications, linear classifiers are not adequate to achieve classification with low error; nonlinear classifiers could be more suitable for such cases.

In the present work, binary nonlinear pattern classification methods are used to perform multiscale and multifeature analysis for the detection of blood vessels. In such a setup, it is desired to classify each pixel in a given image as being a vessel or a nonvessel pixel based on a set of responses of Gabor filters at multiple scales or several features that are extracted from the image and characterize information regarding the blood vessels in different manners. The following sections describe two pattern classification methods that are used for this purpose in the present work.

3.4.1 MULTILAYER NEURAL NETWORKS

Multilayer neural networks (MNN) exploit the same concepts as linear discriminant functions, but provide the optimal solution to an arbitrary problem by implementing linear discriminants in a space where the inputs have been mapped nonlinearly; the parameters of the nonlinearity can be learned from a training set of data at the same time as the parameters that govern the linear discriminant function [167]. Backpropagation, which is based on gradient descent in error (an extension of the least-mean-squares algorithm), is one of the most popular methods for supervised training of the MNN.

MNN are made up of at least three layers [167]. The first layer is the input layer, the size of which is dictated by the number of the features, m_f, available in the feature matrix. The second layer is called the hidden layer, which has a user-specified number of hidden units, n_h. The final layer is called the output layer; the number of nodes in the output layer is the same as the dimensionality of the target vector used for training the MNN. In the present work, a given pixel in the image belongs to only one of two classes; hence, the output layer of an MNN has only one node throughout the present work. The function of such a network is loosely based on the properties of biological neurons; hence, the name artificial neural networks is used to describe such networks.

Assume that a feature matrix of size $m_f \times n_s$ is applied to the input layer of an MNN, where n_s is the number of available training samples (number of training pixels in the present work) per feature. The sole purpose of the input layer is to select an $m_f \times 1$ vector, one sample at a time, until a specified maximum number of iterations is reached, a specified minimum error is reached, or some other preset condition is met. At each iteration, the output of each input unit is passed on to each hidden unit in the hidden layer. The hidden units compute a weighted sum of their inputs (the inner product of the input vector with the weight vector at each hidden unit) to form scalar net activations, denoted by net in the present work; the net activation at the jth hidden unit of a three-layer MNN is given as

$$net_j = \sum_{i=1}^{m_f} x_i w_{ji} + w_{j0} = \sum_{i=0}^{m_f} x_i w_{ji}, \qquad (3.21)$$

where the subscripts i and j index the units of the input and the hidden layers, respectively; x_i denotes the ith feature (input) and w_{ji} denotes the input-to-hidden-layer weight of the jth hidden unit with respect to the ith input. The values of such weights are referred to as synaptic weights in analogy to neurobiology. It should be noted that, for the sake of simplicity, the input vector, $\mathbf{x} = x_1, x_2, ..., x_{m_f}$, and the weight vector, $\mathbf{w}_j = w_{j1}, w_{j2}, ..., w_{jm_f}$, are augmented by appending a feature value $x_0 = 1$ and a weight w_{j0} to the input and the weight vectors, respectively. The output of the jth hidden unit is a nonlinear function of its net activation input net_j:

$$y_j = f(net_j), \qquad (3.22)$$

where $f(.)$ is called the activation function. The choice of a proper activation function is crucial to the pattern classification problem, as there are certain desired properties that $f(.)$ needs to possess. First, it is desired that the activation function be nonlinear; second, $f(.)$ needs to be continuous, so that $f'(.)$ is defined throughout the specified range of the function. The third desired property of $f(.)$ is that it saturates, that is, it has a maximum and a minimum; this ensures that the weights and the net activations are bounded. Finally, it is convenient if $f(.)$ is monotonic; thus, $f'(.)$ has the same sign throughout the specified range of the function. A tangent sigmoid (tansig) function has all of the properties mentioned above and is used throughout the present work as the activation function of the hidden units of the MNN. The output of the jth hidden unit of a three-layer MNN in terms of its net activation input, as applied to a tansig function, is given as

$$f(net_j) = \text{tansig}(net_j) \rightarrow \begin{cases} 1, & \text{if } net_j \geq 0 \\ -1, & \text{if } net_j < 0. \end{cases} \qquad (3.23)$$

The output units use the outputs of the hidden layer in a similar fashion as the hidden units use the outputs of the input layer; the net activation of the kth output unit is given in terms of the outputs of the hidden layer as

$$net_k = \sum_{j=1}^{n_h} y_j w_{kj} + w_{k0} = \sum_{j=0}^{n_h} y_j w_{kj}, \qquad (3.24)$$

where n_h is the number of the hidden-layer units, and k indexes the units of the output layer; w_{kj} denotes the hidden-to-output-layer weight at the kth output unit with respect to the jth hidden unit. The bias w_{k0} is treated as being equivalent to one of the hidden units whose output is always $y_0 = 1$. The output of the kth output unit is computed as

$$z_k \equiv g(net_k) = g\left(\sum_{j=0}^{n_h} f\left(\sum_{i=0}^{m_f} x_i w_{ji} \right) w_{kj} \right), \tag{3.25}$$

where $g(.)$ could either be a linear or a nonlinear function, such as a tansig. The outputs of the MNN are used along with known (target) values of each input to train the MNN using the backpropagation method to classify similar features.

The backpropagation method works similar to the least-mean-squares algorithm; it minimizes the squared-difference between the network output and the target output by using the gradient descent approach. The power of the backpropagation method is that it allows for both the input-to-hidden-layer and hidden-to-output-layer weight vectors to be learned at the same time.

The number of hidden layer units, n_h, is crucial to proper training of the MNN. If too many hidden units are used, the network gets tuned to a specific training set and may over-fit the test set. On the other hand, if too few hidden units are used, the network may not have enough free parameters to fit the training data and may under-fit the test set. The numbers of hidden units needed for different applications are usually determined empirically. The weights in each layer are initialized randomly from a single distribution to ensure uniform learning.

MNNs are used in the present work to perform multifeature and multiscale analysis in the detection of blood vessels, as explained in Chapter 5.

3.4.2   RADIAL BASIS FUNCTIONS

RBFs are specific types of MNN, where a single hidden layer is adequate for learning complex patterns [167]. Furthermore, RBFs usually have output units with linear functions, have no input-to-hidden-layer weight vectors, and the activation functions of the hidden units are Gaussian functions [167]. The result of the kth output unit of an RBF is computed as

$$z_k^{RBF}(\mathbf{x}) = \sum_{j=0}^{n_h} \phi_j(\mathbf{x}) w_{kj}, \tag{3.26}$$

where n_h is the number of hidden units, $\mathbf{x} = x_1, x_2, ..., x_{m_f}$ is the input vector, and w_{kj} is the hidden-to-output-layer weight at the kth output unit with respect to the jth hidden unit. $\phi_j(\mathbf{x})$ is the output of the jth hidden unit and is given as

$$\phi_j(\mathbf{x}) = \exp[-\beta \|\mathbf{x} - \mathbf{c}_j\|^2], \quad \text{for } \beta > 0, \tag{3.27}$$

for a Gaussian activation function, where $\| \ \|$ is the Euclidean norm and $\mathbf{c}_j$ is the center vector of the jth hidden unit (Gaussian activation function). The same training methods as for the MNN are used to train the RBF. A large spread value (β) for the Gaussian function can cause under-classification by combining multiple classes into one, whereas a small spread value can cause over-classification by dividing one class into multiple classes when using the RBF.

As previously mentioned, the number of output layer units is always unity ($k = 1$) in the present work. RBF networks are used in the present work to perform multiscale and multifeature analysis in the detection of blood vessels, as described in Chapter 5.

3.5   REMARKS

In this chapter, methods for the detection of oriented patterns in images were described, including Gabor filters and coherence. The GHT was presented as a method for the detection of parametric curves, such as parabolas. Pattern classification methods were described as the means to perform multiscale and multifeature analysis for the detection of blood vessels. In the present work, Gabor filters are used for single-scale detection and analysis of blood vessels, as described in Section 5.3. Multiscale analysis is performed in the present work using multiple scales of Gabor filters with pattern classification methods, as described in Section 5.4. Multifeature analysis is performed using the Gabor magnitude response, coherence, and the inverted green component images, as described in Section 5.5. Detection of the MTA using the GHT is described in Chapter 6.

CHAPTER 4

Databases of Retinal Images and the Experimental Setup

This chapter gives an overview of the DRIVE [168], the STARE [169], and the Telemedicine for ROP in Calgary (TROPIC) databases. The DRIVE database is used in this book to test and evaluate the performance of the presented methods. The STARE and TROPIC databases are used to demonstrate potential clinical applications of the presented methods. A few other databases of retinal fundus images are also reviewed in this chapter. Details of the methods used to evaluate the results of detection and modeling of retinal vasculature are also discussed.

4.1 DATABASES OF RETINAL IMAGES

4.1.1 THE DRIVE DATABASE

The images in the DRIVE database were obtained through a screening program for DR in the Netherlands [14]. There were a total of 453 subjects in the screening program between the ages of 31 and 86 years, including 400 diabetic subjects from 25 to 90 years [168]. The images were captured using a Canon CR5 nonmydriatic camera with three charge-coupled-device detectors and a field of view (FOV) of 45° [168]. It should be noted that the mapping of the approximately spherical surface of the back of the eye (the retina) onto a rectangular image frame could cause a small amount of geometrical distortion and uneven spatial resolution.

Each image in the DRIVE database was acquired at 8 bits per red, green, and blue (RGB) plane of size 768×584 pixels. However, the images were cropped to the size of width 565 and height 584 pixels, because the FOV has a diameter of about 540 pixels [168]. In the present work, the convention used to state the size of images or indicate coordinate points specifies the row coordinates (annotated by m) first, followed by the column coordinates (annotated by n); hence, the size of the images of the DRIVE database is stated to be 584×565 pixels. The DRIVE images are considered to be low-resolution fundus images of the retina; considering the size of the FOV, the images have an approximate spatial resolution of 20 μm per pixel. The images were compressed using the Joint Photographic Experts Group (JPEG) format and were saved using Tagged Image File Format (TIFF) handles.

Forty images were selected at random to be included in the DRIVE database, including seven showing signs of mild DR, such as exudates, hemorrhages, and pigment epithelium changes. Thirty-three images do not show any abnormal signs. The set of 40 images has been divided into a training set and a test set, each containing 20 images. One manually segmented image of the vasculature is available for each image in the training set, whereas the test set has two manually segmented images of the vasculature per image. The observers who manually marked the vasculature were all trained and instructed by an expert ophthalmologist. The observers were asked to mark a pixel as being part of a vessel only if they were at least 70% certain that the given pixel belonged to the vessel [168]. A mask image is also included in the database to indicate the FOV for each of the 40 images. However, these masks were not used in the present work; instead, new masks were generated using the procedure described in Section 5.2 to avoid edge artifacts.

4.1.2   THE STARE DATABASE

The STARE database [169] is one of the first and oldest publicly available databases of retinal images. The images of the STARE database are scanned and digitized color fundus photographs [170] and, as a result, the quality of the images is not as good as retinal images that are directly captured in a digital format. The images of the STARE database have a narrow FOV of 35° and a size of 700 × 605 pixels. The spatial resolution of the images is approximately 15 μm/pixel; however, the spatial resolution appears to vary from one image to another and is not constant.

For the purpose of evaluation of the diagnostic performance of the detection and modeling methods in a potential clinical application (see Chapter 7), 11 normal cases and 11 cases of PDR were obtained from the STARE database. The STARE database has a total of 22 cases that are diagnosed with PDR; however, 11 cases were not used because either the MTA was not in the FOV, or the major nasal/temporal branches were not distinguishable. The 11 images of normal cases were selected starting from the lowest image number in the subset of images of the STARE database that is used for the detection of the ONH; cases that either did not clearly show the entire MTA, or possessed poor contrast within the FOV, were not selected.

The STARE database is the only publicly available database that includes cases of PDR. Previously, the PDR cases were not part of the publicly available subset of images of the STARE database and were provided by Dr. A. Hoover (Clemson University) upon request; all 397 images of the STARE database are now publicly available [169].

4.1.3   THE TROPIC DATABASE

The diagnostic performance of the modeling methods presented in this book is also evaluated using a private database of retinal fundus images of preterm infants called the TROPIC database [171]. The images of the TROPIC database were captured using the wide-field (130°) RetCam 130 camera and have a size of 640 × 480 pixels. The spatial resolution of the RetCam 130 images is estimated to be 30 μm per pixel [172].

In total, 110 images from 41 patients were randomly selected from the database for the proposed clinical application (see Chapter 7). Ninety of the 110 images were diagnosed with no ROP or stage 1 or 2 ROP (30 images per category), and 20 images were diagnosed with stage 3 ROP. Nineteen of the 110 selected images are from patients diagnosed with plus disease (stages 2 and 3 of ROP), and 91 show no signs of plus disease. At most, two images from the same patient were included for the same stage of ROP (one image from each eye). Images of the same eye from the same patient were included only if the ROP stages were different at the time of imaging, as diagnosed by an expert retinal specialist (A. L. Ells).

4.1.4 OTHER DATABASES

In recent years, there has been an increase in the number of publicly available databases of retinal images of diabetic cases. These databases include, DIARETDB1 [173], MESSIDOR [174], ROC [175], and HEI-MED [176]. All of the mentioned databases only include cases of NPDR, with the exception of HEI-MED, which includes cases of diabetic macular edema. These databases were not used in this work and are only included for the reader's reference.

4.2   ANNOTATION OF IMAGES OF THE DRIVE DATABASE

To evaluate the performance of the methods for the detection of retinal vessels using Gabor filters, only one of the two sets of manual segmentation provided with the DRIVE test set was used, which shall be referred to as the ground-truth images. However, the DRIVE database does not provide separate manual marking of only the MTA. Hence, for the purpose of evaluation of the performance of the proposed modeling methods, the STAs and ITAs in all of the 40 DRIVE images were traced by an expert ophthalmologist and retinal specialist (Dr. Anna L. Ells, Alberta Children's Hospital, Calgary), by magnifying the original image by 145% using the software ImageJ [177]. Only the main venule (the thickest branch of the vessels) was traced within the FOV. At each branching point, the thicker branch was followed [see Figure 4.1 (c)]. The availability of separate traces of the STA and the ITA facilitates the assessment of the accuracy of the dual-parabolic modeling procedure, as described in Section 6.7. The hand-drawn traces of the STA and the ITA can be combined to obtain the trace of the MTA.

The positions of the fovea in all of the 40 DRIVE images were also marked, using the same setup, by Dr. Ells. The centers of the ONH in all of the images were also marked by Dr. Ells, as described in an earlier report [116]. When labeling the center of the ONH, care was taken not to mark the center of the optic cup or the focal point of divergence of the central retinal venule and arteriole [see Figure 4.1 (b)]. In the present work, the manual markings of the ONH center and the fovea were used to correct for any rotation existing between the retinal raphe and the horizontal axis of the image (see Figure 1.6).

4.3   METHODS FOR EVALUATION OF THE RESULTS

There are two separate methods in the present work that need to be tested for quantitative analysis of their performance: the detection of the vascular architecture using Gabor filters (Section 3.2.2), and the parametric modeling of the MTA using the GHT (Section 3.3.2). ROC analysis is used to assess the accuracy and efficiency of the vessel-detection algorithm. The procedure for parabolic

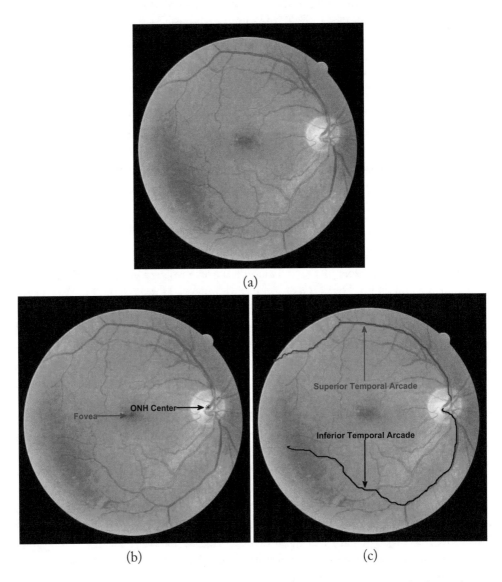

(a)

(b)   (c)

Figure 4.1: (a) Image 17 of the DRIVE database (584 × 565 pixels); this is a standard macula-centered image. (b) The same image as in (a) showing the fovea and the center of the ONH as marked by Dr. Ells; the "*" mark denotes the location of the center of the ONH identified by the ophthalmologist. (c) The same image highlighting the arcades as traced by Dr. Ells; the superior and inferior parts of the arcade were drawn separately and are shown in different colors.

modeling of the MTA relies on a VST, which ideally, should only contain the MTA. In order to measure the effects of artifacts, such as the presence of thicker arterioles and unremoved smaller vessels, on the parabolic modeling of the MTA, the modeling procedure is applied to both the automatically obtained VSTs (*Auto*) and the hand-drawn traces of the MTA (*Hand*). The parameters a of the two models are compared using correlation coefficients. The Euclidean distance between the two detected vertices is used as a distance-error measure. To assess the accuracy of the parabolic models as compared to the hand-drawn traces of the MTA, the mean distance to the closest point (MDCP) and the Hausdorff measures are obtained as error values. The following sections provide details of the performance measures mentioned above.

4.3.1   RECEIVER OPERATING CHARACTERISTICS

In the present work, ROC analysis is used to evaluate the accuracy of the images of detected blood vessels obtained using methods for single-scale, multiscale, and multifeature analysis, as described in Chapter 5. ROC analysis is also used to select an optimal threshold to binarize the grayscale images produced by the methods for the detection of blood vessels. A pixel in a ground-truth image can belong to a vessel (white), which shall be referred to as a positive decision, also indicated by the positive value (1), of the pixel. Similarly, a nonvessel (black) pixel shall be referred to as a negative decision, having a value of 0. According to this convention, a pixel in the detected vessel map can assume one of four different labels. If it is labeled as a vessel pixel by a proposed method, and is also marked as a positive pixel in the corresponding ground-truth image, then it is labeled as true positive (TP). If the pixel is labeled as a vessel in the obtained vessel map, but it is marked as nonvessel in the ground-truth image, then it is labeled as false positive (FP). By the same convention, the labels of true negative (TN) and false negative (FN) are also defined for pixels that are identified as nonvessel pixels in the obtained vessel map. For the sake of normalization, two additional quantities are defined: the total number of actually positive (AP) pixels or cases, and the total number of actually negative (AN) pixels or cases. It is important to note that, in order to measure the performance of a diagnostic decision, prior knowledge is required regarding the number of AP and AN cases that are present in the whole database.

When performing an evaluation of medical diagnostic decisions, simply stating the accuracy of the diagnosis may be insufficient and misleading. For example, if there is a test population in which 7% of patients have a certain disease, one can perform the diagnosis with 93% accuracy if one were to blindly guess that all patients do not have the disease, that is, all are negative cases. This example demonstrates the need for a more meaningful representation of the diagnostic results. ROC analysis can provide a better understanding of the decision performance by introducing two indices [178]:

$$\text{Sensitivity} = \frac{\# \text{ TP Decisions}}{\# \text{ AP Cases}}, \qquad (4.1)$$

and

$$\text{Specificity} = \frac{\#\,\text{TN Decisions}}{\#\,\text{AN Cases}}. \tag{4.2}$$

Sensitivity is also referred to as the true-positive fraction (TPF); similarly, specificity is referred to as the true-negative fraction (TNF). Two other related terms, which are the false-positive fraction (FPF) and the false-negative fraction (FNF), are defined as follows:

$$\text{FPF} = \frac{\#\,\text{FP Decisions}}{\#\,\text{AN Cases}}, \tag{4.3}$$

and

$$\text{FNF} = \frac{\#\,\text{FN Decisions}}{\#\,\text{AP Cases}}. \tag{4.4}$$

It can be easily shown that $\text{TPF} + \text{FNF} = 1$:

$$\text{TPF} + \text{FNF} = \frac{\#\,\text{TP Decisions} + \#\,\text{FN Decisions}}{\#\,\text{AP Cases}},$$

but FN decisions are, in fact, positive cases; hence, $\text{TP} + \text{FN} = \text{AP}$, and $\text{TPF} + \text{FNF} = 1$. Similarly, $\text{TNF} + \text{FPF} = 1$. These equations can also be rearranged using sensitivity and specificity as $\text{FPF} = 1-$ specificity and $\text{FNF} = 1-$ sensitivity.

In order to determine the actual values of the fractions and measures of performance described above, it is necessary to have a threshold to divide the results into the two categories of positive and negative cases. All cases below the threshold are diagnosed as being negative and cases above the threshold are diagnosed as the opposite, or vice versa. After dividing the results into two categories, they are compared with the corresponding ground-truth values and the FPF and TPF values are calculated. By sorting the detected values and the corresponding ground-truth values, and applying a sliding threshold varying from the lowest value to the highest, one may analyze the trade-off between the TPF and the FPF as the threshold changes; this constitutes ROC analysis and can help in finding an optimal threshold for binarizing the continuous-valued decision variable. Furthermore, ROC analysis can also help by providing a measure of performance for the method under review. By varying the threshold and plotting the resulting TPF (sensitivity) values against the FPF ($1-$ specificity) values, the ROC curve is obtained; the area under the ROC curve (A_z) indicates the accuracy of the obtained results, based on the varying threshold. The point on the ROC curve that has the shortest distance to the point $(0, 1)$ can be taken as an optimal threshold to binarize the decision variable. A 100% accurate test has $A_z = 1$ represented by a square ROC curve.

4.3.2   DISTANCE MEASURES

Euclidean Distance

Given two detected vertices, one being used as the reference, $(x_{o_{Hand}}, y_{o_{Hand}})$, and the other being treated as its detected estimate, $(x_{o_{Auto}}, y_{o_{Auto}})$, one can use the Euclidean distance between the two points as a distance or error measure to determine the accuracy of detection. The Euclidean distance is given as [179]

$$d = \sqrt{(x_{o_{Hand}} - x_{o_{Auto}})^2 + (y_{o_{Hand}} - y_{o_{Auto}})^2}. \qquad (4.5)$$

The smaller the Euclidean distance between the two vertices, the closer they are together, and hence, the more accurate the estimate is. The Euclidean distance between the vertices of the parabolic fits to the hand-drawn traces of the MTA and the corresponding automatically detected MTA is calculated for each of the 40 images in the DRIVE database. The mean and the STD of the Euclidean distances for all 40 images are then obtained.

Mean Distance to the Closest Point

The MDCP measures the closeness of two given contours based on the mean of the distance to the closest point (DCP) from one of the contours (the model) to the other (the reference). Given a model, $M = \{m_1, m_2, ..., m_N\}$ with N points, and a reference, $R = \{r_1, r_2, ..., r_K\}$ with K points, the DCP for a single point m_i on M is defined as [180]

$$\text{DCP}(m_i, R) = \min(\|m_i - r_j\|), \ j = 1, \ 2, \ ..., \ K, \qquad (4.6)$$

where $\| \ \|$ is a norm operator, such as the Euclidean norm. The MDCP measure from M to R is computed as

$$\text{MDCP}(M, R) = \frac{1}{N} \sum_{i=1}^{N} \text{DCP}(m_i, R). \qquad (4.7)$$

Figure 4.2 illustrates the DCP errors as measured between two contours; the green contour is taken as the reference and the red contour is taken as the result of modeling. The DCP, from the model to the reference, is shown in the illustration only for every fifth point on the result of modeling.

In the present work, the MDCP is measured from the parabolic model, obtained using each of the different variations of the GHTs and the dual-parabolic modeling procedure, to the corresponding hand-drawn trace of the MTA (see Section 6.8.2). The smaller the MDCP value, the closer the model matches the reference. The MDCP is also used in the MTA detection process to select the parameters of the best-fitting parabolic model among the top 10 global maximum candidates in the Hough space, as explained in Section 6.5.

Figure 4.2: An image, of size 584×565 pixels, illustrating the DCP and the MDCP. The green parabolic contour is considered to be the reference, given by the parameters $(270, 125, 55)$, where the first two values indicate the vertex location, (m_o, n_o), and the third value is the openness parameter (a), as defined in Equation 3.20. The red parabolic contour is considered to be the model with the parameters $(285, 145, 65)$. The cyan lines connecting the model to the reference show the DCP values; the DCP is shown only for every fifth point on the model. The MDCP for this example was calculated to be 16.85 pixels.

The Hausdorff Measure

The Hausdorff measure [181] indicates the longest distance from the model to the reference. The Hausdorff measure is comparable to the MDCP as it first finds the DCP from each point on the model to the reference. However, instead of taking the mean of all the DCPs, the Hausdorff measure takes their maximum. Given a model, $M = \{m_1, m_2, ..., m_N\}$, and a reference, $R =$

$\{r_1, r_2, \dots, r_K\}$, the DCP for a single point, m_i, on M is derived using Equation 4.6. The Hausdorff measure, H, from M to R is computed as

$$H(M, R) = \max\{DCP(M, R)\},\qquad(4.8)$$

where $DCP(M, R)$ is an $N \times 1$ vector representing the DCPs for each point on M, computed using Equation 4.6. It should be noted that the Hausdorff measure is not symmetric, that is, $H(M, R) \neq H(R, M)$.

In the present work, the Hausdorff measure is used as an error measure to assess the accuracy of the parabolic models of the MTA, as described in Section 6.8.2.

4.3.3   CORRELATION COEFFICIENT

The correlation coefficient is essentially the normalized covariance between two variables. The covariance measures how two variables, for example, a_{Hand} and a_{Auto}, vary jointly. The covariance is calculated as [182]

$$C(a_{Hand}, a_{Auto}) = \frac{\sum (a_{Hand} - \mu_{a_{Hand}})(a_{Auto} - \mu_{a_{Auto}})}{n},\qquad(4.9)$$

where μ indicates the mean of the population, n is the number of available samples, and the summation is done over all available samples.

The correlation coefficient, r, is the normalized covariance value obtained as [182]

$$r = \frac{C(a_{Hand}, a_{Auto})}{\sqrt{C(a_{Hand}, a_{Hand}) C(a_{Auto}, a_{Auto})}};\qquad(4.10)$$

however, the covariance of the same feature is its variance, that is [182],

$$C(a_{Hand}, a_{Hand}) = \sigma^2_{a_{Hand}} \text{ and } C(a_{Auto}, a_{Auto}) = \sigma^2_{a_{Auto}},$$

where σ represents the STD; hence Equation 4.10 can be written as [182]

$$r = \frac{C(a_{Hand}, a_{Auto})}{\sigma_{a_{Hand}} \sigma_{a_{Auto}}}.\qquad(4.11)$$

A correlation coefficient close to unity $(+1)$ indicates that either variable is predictable from the other and that they both increase together. A correlation coefficient close to zero means that neither variable provides information regarding the other.

In the present work, the correlation coefficient between two sets of a parameters is used to compare the results of parabolic modeling of the MTA, as explained in Section 6.8. The correlation coefficient between the automatically obtained values of the parameter a and the TAA, measured as described by Wilson et al. [40], is also obtained, as described in Section 6.10.

4.4   REMARKS

This chapter provided details about the DRIVE database, which is used to test and evaluate the performance of methods for the detection of blood vessels and parabolic modeling of the MTA. An ophthalmologist marked the center of the ONH and the fovea in all of the images; the DRIVE database does not provide such information. The same ophthalmologist also traced the MTA in each of the 40 images of the DRIVE database to facilitate evaluation of the results of parabolic modeling of the MTA. The methods used for the evaluation of the results, including ROC analysis, distance measures, and the correlation coefficient, were also described in this chapter. The results of vessel detection and parabolic modeling, obtained using the DRIVE database, and their corresponding evaluation, are presented in Chapters 5 and 6, respectively.

CHAPTER 5

Detection of Retinal Vasculature

As mentioned in Section 3.2.2, real Gabor filters are utilized in the present work for the detection of blood vessels in retinal fundus images. Analysis of the color components of the retinal fundus images is a crucial step in providing a grayscale image that attains the highest contrast for the blood vessels for Gabor filtering. An analysis of the color components of the retinal images in the DRIVE database is presented in Section 5.1. In order to avoid the detection of the edges of the FOV by the Gabor filters, appropriate preprocessing steps need to be included, as explained in Section 5.2.

Blood vessels in the retina vary in thickness in the range 50–200 μm, with a median of 60 μm [5, 7]. In a comparative analysis of the performance of the Gabor filter and other line detectors [144, 145], the capture range of a given Gabor filter with the parameter τ, in terms of detecting lines with an efficiency of more than 90%, in the presence of noise with the normalized STD of 0.2, was determined to be about 0.4τ to 3.2τ. This result implies the adequacy of a single-scale Gabor filter to detect blood vessels in the range mentioned above; results of single-scale filtering and analysis of blood vessels, as applied to the images of the DRIVE database, are provided in Section 5.3. Although single-scale analysis could be adequate for high detection rates, it might be beneficial to apply Gabor filters at different scales for multiscale analysis using pattern classification methods; the results of multiscale filtering and analysis of blood vessels, as applied to the images of the DRIVE database, are provided in Section 5.4.

Coherence, as described in Section 3.2.3, could provide additional information regarding blood vessels by taking into account the Gabor angle response; furthermore, the green component of the *RGB* color space [183] could also provide additional information in terms of high vessel-to-background contrast that could be used in the classification of blood vessels. Results of multifeature analysis of the images in the DRIVE database using different Gabor filter scales, coherence, and the green component images, performed using pattern classification methods, are presented in Section 5.5. A comparative analysis of the methods and results presented in the current work with similar methods and results, as given in other publications, is provided in Section 5.6.

Preliminary investigations related to the present work have been reported by Rangayyan et al. [86, 146, 148] and Oloumi et al. [147]. The present work, as described in this chapter, includes further studies on multiscale analysis by including more scale factors, τ, and also by changing the structure of the classifiers used. The present work also includes new studies on multifeature analysis using the Gabor magnitude response, coherence, and the inverted G component images.

5.1   ANALYSIS OF COLOR IMAGE COMPONENTS

As stated in Section 3.2.2, the real Gabor filters are sensitive to oriented features that have a positive contrast against the background. The input image for Gabor filtering is represented in grayscale, where each pixel has a value between 0 and 255 in an 8-bit representation, or 0 and 1 if normalized (with 64-bit representation in the present work). However, the images in the DRIVE database are color images represented using the RGB color space [183, 184]. Each color image in the DRIVE database is composed of three scalar-valued images of size 584×565 pixels, one for each of the red, green, and blue components. For the purpose of detection of blood vessels, each RGB component in the color images of the DRIVE database is analyzed in terms of vessel-to-background contrast.

Figure 5.1 shows the original image 6 of the DRIVE database in color and its corresponding color components, in their respective colors, in order to demonstrate an easily comprehensible representation of each color component. As seen in Figures 5.1 (b) and (d), the R and the B components appear to be noisy and possess a low vessel-to-background contrast, in particular the B component. However, there still may be useful information in the R and the B components. Due to this reason, the luminance component Y of the YIQ color space [183], which computes the luminance of a color image, is also analyzed in terms of vessel-to-background contrast. The YIQ color space is used in the North American TV broadcasting systems to encode color information in terms of luminance (the Y component), as well as the I and Q components, which jointly define hue and saturation, respectively. The YIQ color space is obtained from the RGB color space using a linear transformation as [183, 184]

$$\begin{bmatrix} Y \\ I \\ Q \end{bmatrix} = \begin{bmatrix} 0.299 & 0.587 & 0.114 \\ 0.596 & -0.274 & -0.322 \\ 0.211 & -0.253 & -0.312 \end{bmatrix} \begin{bmatrix} R \\ G \\ B \end{bmatrix}. \tag{5.1}$$

The Y component ($Y = 0.299R + 0.587G + 0.114B$) puts more emphasis on the G component of the RGB color space than its R and B components combined; the Y component is known to be a suitable candidate for the detection of edges in color images [183].

Figure 5.2 shows the grayscale versions for each RGB component as well as the Y component for image 6 of the DRIVE database; it can be observed that blood vessels are darker than their immediate background, that is, blood vessels have negative contrast in all of the images shown. Figure 5.3 shows the inverted versions of the images in Figure 5.2, achieved by subtracting the grayscale images from 255. Such a representation provides positive contrast for the blood vessels, which is required by the real Gabor filters.

Several of the works and methods reviewed in Section 2.1.1 used the G component for analysis, because it provides the best vessel-to-background contrast among the three RGB color

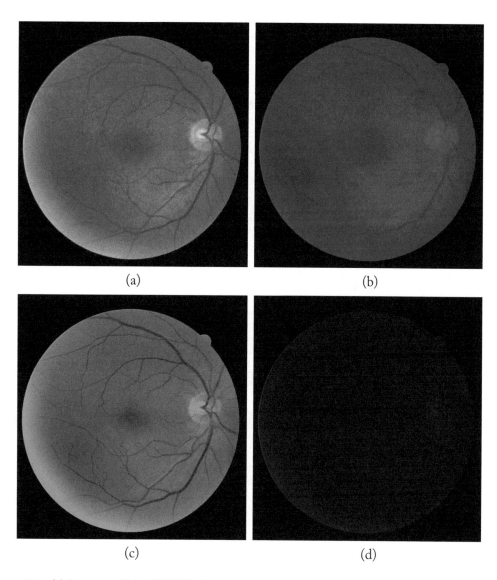

Figure 5.1: (a) Image 6 of the DRIVE database. Each *RGB* color component of the image in (a) is shown in its respective color: (b) the *R* component shown in red, (c) the *G* component in green, and (d) the *B* component in blue. The blood vessels are clearly visible in (c), partially visible in (b), and barely visible in (d).

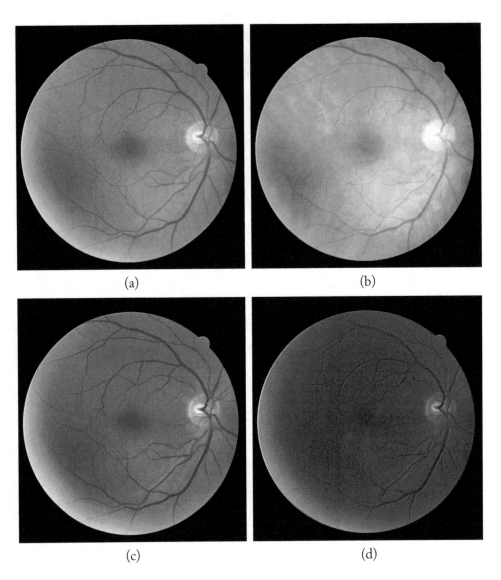

(a)   (b)

(c)   (d)

Figure 5.2: Grayscale representations of image 6 of the DRIVE database linearly mapped to the full range of display for all images: (a) the Y component of the YIQ color space, and (b) the R, (c) the G, and (d) the B components of the RGB color space. It should be noted that blood vessels have negative contrast in all of the images shown.

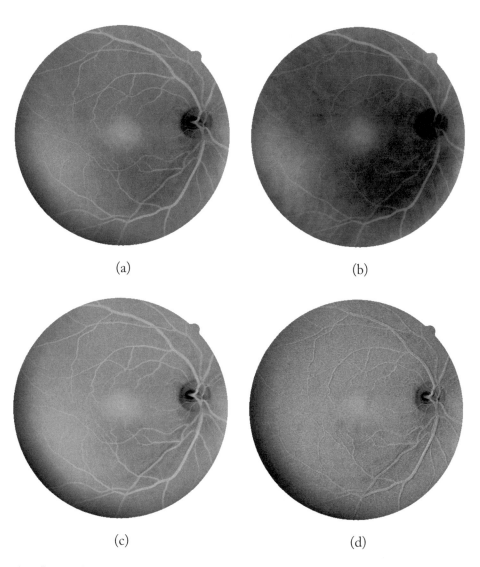

(a)   (b)

(c)   (d)

Figure 5.3: Inverted grayscale representations of the images shown in Figure 5.2; all images are linearly mapped to the full range of display: (a) the inverted Y, (b) the inverted R, (c) the inverted G, and (d) the inverted B components. Note that blood vessels have positive contrast in all of the images shown.

components. The Y component also shows a high vessel-to-background contrast, similar to the G component. In the present work, single-scale analysis is performed using both the Y and the G components and the results are compared in terms of the area under the ROC curve. The Y component is used with Gabor filters in the present work for multiscale and multifeature analysis because it provides marginally better results as compared to the G component (see Section 5.3) [86].

5.2   PREPROCESSING OF RETINAL FUNDUS IMAGES

Figure 5.4 illustrates the steps of the preprocessing module using a flowchart representation; the preprocessing routine is used in the present work to prepare the original color images of the DRIVE database for the application of Gabor filters. The given color image is normalized first to provide a better computational representation of each pixel with a 64-bit representation of values in the range [0, 1] for each component, as compared to the original 8-bit representation. The Y component is obtained using the appropriate part of Equation 5.1 and is then thresholded at 0.1 of the highest intensity value to generate a mask. The thresholding process assigns a value of 1 to each pixel in the Y component that has a value that is more than 0.1; a value of 0 is assigned to each pixel that has a value less than 0.1. All of the pixels in the FOV of the Y component of the images of the DRIVE database have an intensity value that is higher than 0.1; however, some pixels outside the FOV may also have an intensity higher than 0.1. To remove such unwanted segments from outside of the FOV, the morphological operation of area open, as explained in Section 3.1.5, is used to remove 8-connected segments of white pixels that have an area of 100,000 pixels or less. Binary erosion, as described in Section 3.1.1, using a disk-shaped SE of radius $r = 5$ pixels, is used to remove the edge artifacts caused by the thresholding step. The resulting mask is passed on as one of the outputs of the preprocessing module and also is used further in the preprocessing procedure.

In order to avoid edge artifacts in the results of Gabor filtering (the edges of the FOV also are oriented patterns), each image is extended beyond the limits of its effective region as follows [85, 146, 147]: first, a four-pixel neighborhood is used to identify the pixels at the outer edge of the effective region; this is achieved by dilating (see Section 3.1.2) the generated mask using a disk-shaped SE of radius $r = 1$ and subtracting the dilated mask from the original undilated mask. For each of the pixels identified, the mean gray level is computed over all pixels in a 21×21 neighborhood that are also within the effective region, as indicated by the original undilated mask. The mean value is assigned to the corresponding pixel location in the grayscale image. The effective region is merged with the outer-edge pixels, forming an extended effective region. The procedure is repeated 50 times, extending the image by a ribbon of pixels of width 50 pixels. This step homogenizes the pixels around the edges of the FOV with the pixels outside the FOV and eliminates sharp transitions (edges).

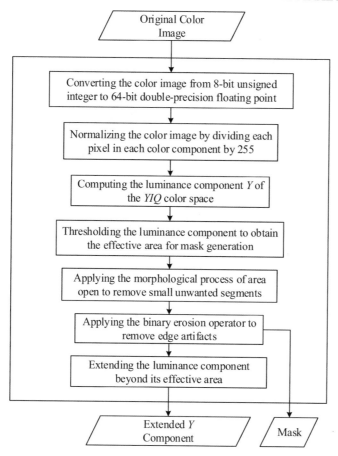

Figure 5.4: Flowchart representation of the preprocessing module. The preprocessing routine takes a color retinal fundus image as the input and returns the extended Y component, as well as the generated mask.

Figure 5.5 shows the result of applying the preprocessing routine to image 6 of the DRIVE database. The inverted and extended Y-component image, as well as the generated mask, are supplied to the Gabor filter module for the detection of blood vessels, as explained in the following section.

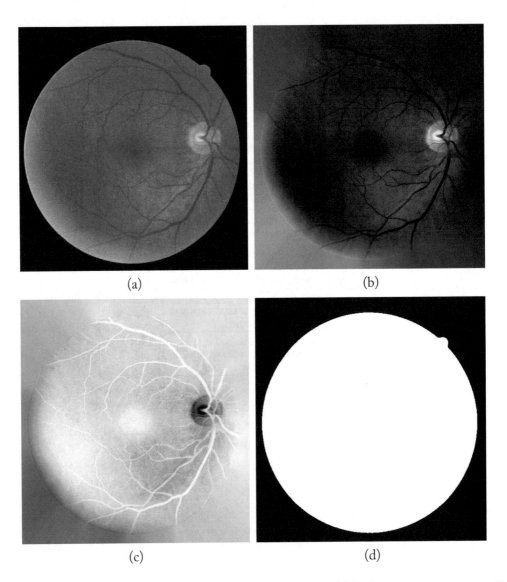

(a)   (b)

(c)   (d)

Figure 5.5: (a) Original color image 6 of the DRIVE database. (b) The luminance component, Y, of (a) extended beyond its FOV by the preprocessing routine to avoid the detection of the edges of the FOV by the Gabor filters. (c) The inverted and extended Y component, to be provided as input to the Gabor filters. (d) The mask generated by the preprocessing routine; the mask is used to restrict analysis to the effective area of the retinal image.

5.3   SINGLE-SCALE FILTERING AND ANALYSIS

Figure 5.6 gives an algorithmic flowchart representation of the Gabor filtering module, as used in the present work. The Gabor filter routine takes, as inputs, the grayscale inverted and extended Y-component image, as well as the binary mask image provided by the preprocessing module. The Gabor filter routine also requires the values for the Gabor filter design parameters τ and l, as well as the preferred number of filters, K, spanning the range $[-\pi/2, \pi/2]$. In the present work, $K = 180$ filters is used for all of the illustrations as well as all of the analysis performed. The main Gabor filter is constructed directly in the Fourier domain in the present work to avoid unnecessary computation associated with the application of the Fourier transform (FT) to the impulse response of the Gabor kernel to obtain its frequency response. The Fourier spectrum of the inverted and extended Y image is obtained using the 2D discrete FT via the fast FT (FFT) algorithm, and shifted such that the DC point, or $(0, 0)$ frequency is at the center; the result is high-pass filtered to remove the DC point before the application of Gabor filters. The resulting magnitude and angle response images are masked to obtain the results only for the effective area of the image.

Figure 5.7 shows the result of applying Gabor filters at different scales to the inverted and extended Y-component image in Figure 5.5 (c). The magnitude-response images are obtained by selecting the maximum response over all 180 Gabor filters for each pixel. The result in Figure 5.7 (a) illustrates that the filters have detected only the edges of the thick vessels, with poor response along their center-lines. On the contrary, the result in Figure 5.7 (b) shows that, while the thick vessels have been detected well, some of the thinner vessels have not been detected. The results of Figure 5.7 (c) show that increasing the elongation parameter of the Gabor filters can cause over-fitting in places where blood vessels bend. These results indicate that some advantage could be gained by performing multiscale filtering and analysis, which is easily facilitated by the proposed design of the Gabor filters (see Section 5.4).

The proposed methods also provide the dominant orientation at each pixel, obtained as the angle of the filter with the largest magnitude response at the same pixel. Figure 5.7 (d) shows the angle data, in the form of needles for every fifth pixel, for a part of the image in Figure 5.2 (a). The angle data exhibit a high level of agreement with the local orientation of the blood vessels.

The parameters of the Gabor filters were varied over the range $[1, 16]$ pixels for τ and $[1.3, 18.5]$ for l. For each set of the parameters $\{\tau, l\}$ used, the magnitude responses of the Gabor filters, as applied to the 20 images in the training set of the DRIVE database, were acquired and masked to obtain the effective regions. ROC analysis (see Section 4.3.1) was applied to the magnitude-response images and the corresponding ground-truth images provided in the DRIVE database to determine the TPF and the FPF. Only the effective region of each image was used to perform ROC analysis. The area under the ROC curve (A_z) was computed as an indicator of the performance of the detection procedure. Figure 5.8 shows the results of thresholding a Gabor magnitude-response image. An optimal threshold was obtained by finding the point on the ROC curve that is the closest to the point $(0, 1)$.

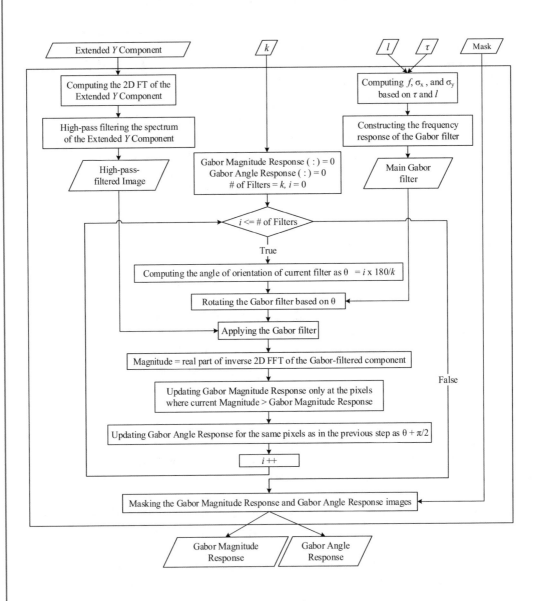

Figure 5.6: Flowchart representation of the Gabor filtering module for single-scale detection of blood vessels.

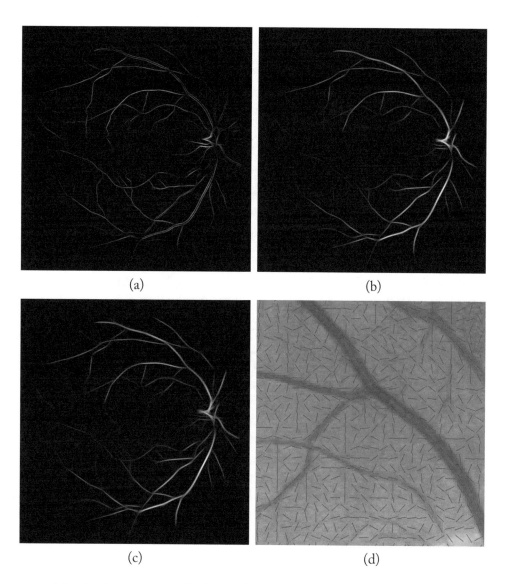

Figure 5.7: Magnitude response of 180 Gabor filters over the range $[-\pi/2, \pi/2]$, as applied to the image in Figure 5.5 (c) with: (a) $\tau = 4$ pixels, $l = 2.9$; (b) $\tau = 8$ pixels, $l = 2.9$; and (c) $\tau = 8$ pixels, $l = 4$. (d) Angles related to (b), shown in the form of needles for every fifth pixel for a part of the image.

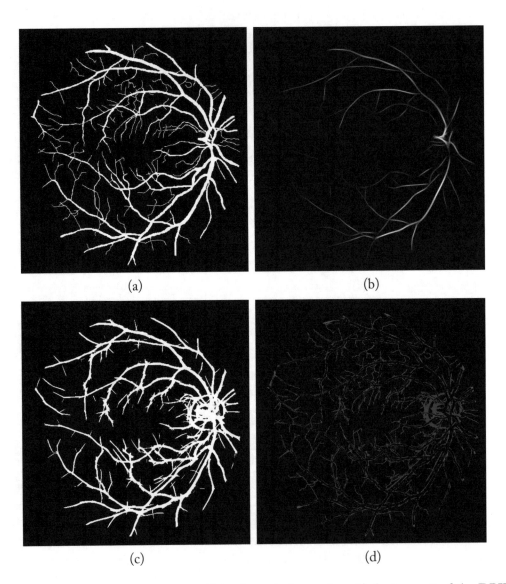

(a)

(b)

(c)

(d)

Figure 5.8: (a) Manual segmentation of blood vessels (ground truth) for image 6 of the DRIVE database shown in Figure 5.5 (a). (b) Magnitude response of the Gabor filters with $\tau = 8$ and $l = 2.9$ with the range of values $[0, 0.0272]$. (c) Result of thresholding the magnitude response in part (b) with threshold $= 0.0023$. (d) FP (blue) and FN (red) pixels in the result (c) as compared to the image in (a).

Table 5.1 presents the A_z values for some of the combinations of the design parameters for the 20 images in the training set of the DRIVE database obtained using the Y component. The highest blood-vessel detection rate of $A_z = 0.94$ was achieved with $\tau = 8$ pixels and $l = 2.9$ (and a few other sets of the parameters). Using the same parameters, a detection efficiency of $A_z = 0.95$ was obtained with the 20 images in the test set. Table 5.2 presents the A_z values for some of the combinations of the parameters for the 20 images in the training set obtained using the G component. The highest blood-vessel detection rate of $A_z = 0.94$ was achieved with $\tau = 8$ and $l = 2.9$ (and some other sets of the parameters) for the training set of 20 images using the G component. Using $\tau = 8$, $l = 2.9$, and the G component, a detection efficiency of $A_z = 0.95$ was obtained with the 20 images in the test set. Figure 5.9 shows four ROC curves obtained with different values of τ, using the Y and the G components. Both cases with $\tau = 8$ have provided similar results using the Y and G components; the case with $\tau = 1$ shows poor performance.

Table 5.1: Efficiency of the detection of blood vessels (A_z) for the training set (20 images) of the DRIVE database, for selected sets of values of τ (in pixels) and l, using the inverted Y component

Parameters	$l = 1.7$	$l = 2.1$	$l = 2.5$	$l = 2.9$	$l = 3.3$	$l = 3.7$
$\tau = 1$	0.62	0.65	0.67	0.70	0.73	0.75
$\tau = 2$	0.69	0.72	0.75	0.77	0.79	0.81
$\tau = 4$	0.85	0.87	0.89	0.90	0.91	0.92
$\tau = 6$	0.91	0.92	0.93	0.94	0.94	0.94
$\tau = 7$	0.92	0.93	0.94	0.94	0.94	0.94
$\tau = 8$	0.93	0.94	0.94	0.94	0.94	0.94
$\tau = 9$	0.94	0.94	0.94	0.94	0.94	0.93
$\tau = 10$	0.94	0.94	0.94	0.93	0.93	0.92
$\tau = 12$	0.93	0.93	0.93	0.92	0.91	0.91

Table 5.2: Efficiency of the detection of blood vessels (A_z) for the training set (20 images) of the DRIVE database for selected sets of values of τ (in pixels) and l, using the inverted G component. The A_z results using the G component are not provided for all of the $\{\tau, l\}$ combinations given in Table 5.1, as they are similar

Parameters	$l = 1.7$	$l = 2.1$	$l = 2.5$	$l = 2.9$	$l = 3.3$	$l = 3.7$
$\tau = 4$	0.85	0.87	0.89	0.90	0.91	0.92
$\tau = 6$	0.91	0.92	0.93	0.93	0.94	0.94
$\tau = 8$	0.93	0.94	0.94	0.94	0.94	0.94
$\tau = 10$	0.94	0.94	0.94	0.93	0.93	0.92
$\tau = 12$	0.93	0.93	0.92	0.92	0.91	0.91

5.4   MULTISCALE FILTERING AND ANALYSIS

As evident from the high A_z values obtained using single-scale Gabor filtering, a single scale of the Gabor filter is adequate for the detection of most of the blood vessels in a given retinal fundus image. However, multiscale analysis performed by combining other scales using pattern classification methods may help in more accurate detection of thin and thick vessels. MNN and RBF classifiers were used in the present work to perform multiscale analysis for the detection of blood vessels. The multiscale analysis was performed with various combinations of different scales, with $\tau \in \{0.5, 1, 2, 4, 8, 12, 16\}$ and $l = 2.9$. Only the data from the effective areas of the images were used for training and testing. The training step was performed using the 20 images from the training set of the DRIVE database.

In the case of the MNN classifiers, only a fraction of the available training data (10% or 15%) were used to train the classifiers, because large amounts of data can exhaust the available system memory during the training process. The training data were selected at random from each image, along with their corresponding ground-truth data; as previously mentioned, only the pixels inside the FOV were considered for training. The number of hidden-layer nodes for the MNN classifiers was varied in the range [10, 60] nodes and the number of hidden layers was varied between one and three layers. A tansig activation function was used for all hidden-layer nodes; the output-layer function was varied between a tansig and a linear function. The trained classifiers were then tested using the 20 images from the test set of the DRIVE database. Table 5.3 provides the results of blood vessel detection in terms of A_z values, using the MNN classifiers for some of the scale combinations tested. The set $\tau = \{4, 8, 12\}$ and a few other scale combinations provided marginal improvement, as compared to the results of single-scale analysis of the blood vessels, with $A_z = 0.96$. Variations in the number of hidden layers, number of hidden-layer nodes, percentage of training data used, and the output-layer function were observed to have minor effects on the results of multiscale analysis using the MNN classifiers.

The RBF classifiers were trained with only 0.125% of the available training data from the 20 images of the training set in the DRIVE database; using higher percentages of training data caused the program to run out of memory. The number of hidden-layer nodes was varied in the range [8, 15] and the spread (STD) of the Gaussian functions was kept constant at $\sigma = 1.2$, which was selected empirically. The same range of scale values was used in the training of the RBF classifiers as for the MNN classifiers. Table 5.4 provides the results of blood vessel detection in terms of A_z values, using the RBF classifiers for some of the scale combinations tested. Similar to the results of the MNN classifiers, $\tau = \{4, 8, 12\}$ and a few other combinations led to marginal improvement as compared to the results of single-scale filtering and analysis of the blood vessels, with $A_z = 0.96$. Variations in the number of hidden nodes and/or percentage of the training data used did not have a significant effect on the results of multiscale analysis using the RBF classifiers.

Figure 5.10 illustrates examples of discriminant images (outputs of the MNN and RBF classifiers) obtained using multiscale analysis. The result in Figure 5.10 (b), with $\tau = \{8, 12\}$, is missing thin vessels as compared to the result in part (a) of the same figure, with $\tau = \{4, 8, 12\}$,

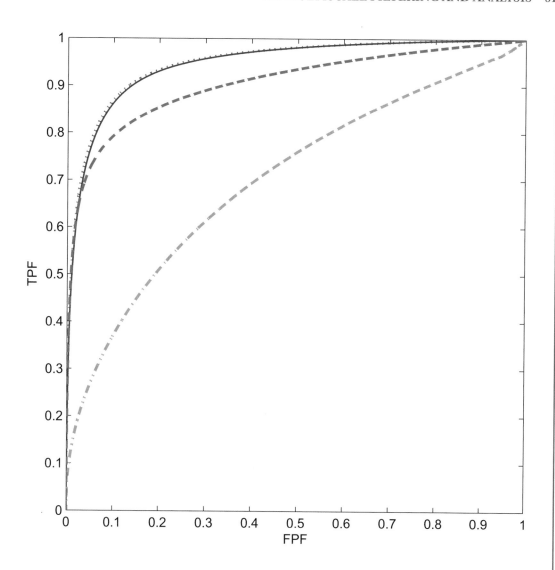

Figure 5.9: Various ROC curves obtained using different values of τ for the test set of 20 images of the DRIVE database; in all cases $l = 2.9$ is used. Dotted red line: $\tau = 8$ using the Y component, with $A_z = 0.9495$. Solid blue line: $\tau = 8$ using the G component, with $A_z = 0.9471$. Dashed magenta line: $\tau = 4$ using the Y component, with $A_z = 0.9071$. Dot-dashed green line: $\tau = 1$ using the Y component, with $A_z = 0.7063$. The x-axis represents 1-specificity (FPF) and the y-axis represents the sensitivity (TPF) values.

Table 5.3: Efficiency of the detection of blood vessels (A_z) with multiscale analysis of the test set (20 images) of the DRIVE database using different MNN structures and scale combinations. The same activation function (tansig) was used in all hidden layers in all cases

Scale $\{\tau\}$	Number of Hidden Layers	Number of Hidden Nodes	Output-Layer Function	Training Data (%)	A_z
$\{4,8\}$	1	20	tansig	15	0.95
$\{4,8\}$	1	20	linear	15	0.95
$\{8,12\}$	1	20	tansig	10	0.95
$\{8,12\}$	1	20	linear	10	0.95
$\{0.5,4,8\}$	1	20	tansig	10	0.95
$\{0.5,8,12\}$	1	20	linear	10	0.95
$\{1,4,8\}$	1	20	tansig	10	0.95
$\{1,4,8\}$	1	20	linear	10	0.95
$\{1,8,12\}$	1	20	tansig	10	0.95
$\{1,8,12\}$	1	20	linear	10	0.95
$\{4,8,12\}$	1	20	tansig	10	0.96
$\{4,8,12\}$	1	10	linear	15	0.96
$\{0.5,4,8,12\}$	1	30	linear	15	0.96
$\{0.5,4,8,12\}$	2	10, 10	tansig	10	0.96
$\{1,4,8,12\}$	1	20	tansig	10	0.96
$\{4,8,12,16\}$	1	30	linear	15	0.96
$\{0.5,1,4,8,12\}$	2	20, 20	tansig	10	0.96

Table 5.4: Efficiency of the detection of blood vessels (A_z) with multiscale analysis of the test set (20 images) of the DRIVE database using different RBF structures and scale combinations. In all instances, the RBF classifier was trained using 0.125% of the available training data

Scale $\{\tau\}$	Number of Hidden Nodes	A_z
{4,8}	8	0.95
{4,8}	15	0.95
{8,12}	8	0.95
{8,12}	15	0.95
{0.5,4,8}	8	0.95
{0.5,8,12}	8	0.95
{1,4,8}	8	0.95
{1,4,8}	15	0.95
{1,8,12}	8	0.95
{1,8,12}	15	0.95
{4,8,12}	8	0.96
{4,8,12}	15	0.96
{0.5,4,8,12}	8	0.96
{0.5,4,8,12}	15	0.96
{1,4,8,12}	8	0.96
{4,8,12,16}	8	0.96
{0.5,1,4,8,12}	8	0.96

because the former was obtained with one fewer scale. The images in Figure 5.10 (c) and (d) show lower discrimination between vessel and nonvessel (background) pixels, as compared to that in Figure 5.10 (a). The A_z values provided in the caption of Figure 5.10 indicate the area under the ROC curve for each single discriminant image and not the entire test set of 20 images.

Figure 5.11 shows the results of thresholding the discriminant images in Figure 5.10. In each case, an optimal threshold was obtained by finding the closest point from $(0, 1)$ to the ROC curve generated, using the 20 images in the test set. The images obtained using an RBF classifier [Figure 5.11 (c) and (d)] appear to have more FP pixels, as compared to the image obtained using an MNN, as shown part (a) of the same figure. Figure 5.11 (b) is missing the thinner vessels, because the discriminant image used to generate it was obtained using only two scales. Figure 5.12 illustrates the FP and FN pixels for the images in Figure 5.11; the results were obtained by comparing the thresholded images to the ground-truth image [see Figure 5.8 (a)]. The result in Figure 5.12 (a) shows fewer FP pixels, as compared to the other images of the same figure.

5.5   MULTIFEATURE ANALYSIS

Multiscale analysis, as performed in the present work, only employs the Gabor magnitude response at different scales of thickness as provided by the Gabor filters. However, Gabor filters, as designed in the present work, also provide the Gabor angle response. As discussed in Section 3.2.3, the coherence image obtained using the Gabor magnitude and angle responses could also provide information regarding the orientation of the blood vessels in retinal images. Coherence images were derived for all of the 40 images in the DRIVE database using Equation 3.18 with the Gabor magnitude and angle responses obtained with $\tau = 8$ and $l = 2.9$. As observed in Section 5.1, the inverted G component of the RGB color space provides high contrast for the blood vessels, which is a desirable characteristic of a feature used for the detection of blood vessels; therefore, the inverted G component was also used as a feature to help improve the results of the classification of blood vessels. The MNN and the RBF classifiers were used to perform multifeature analysis for the detection of blood vessels. Similar to the setup for multiscale analysis, only the data from the effective area of the images were used for training and testing.

Table 5.5 presents the results of multifeature analysis for some combinations of scales, coherence, and the inverted G component, in terms of the A_z value obtained using the MNN classifier. The combination of Gabor magnitude responses with $\tau = \{4, 8, 12\}$ and the inverted G component, as well as several other combinations, led to marginal improvements in terms of the area under the ROC curve, as compared to the results of single-scale analysis. Table 5.6 provides the results of multifeature analysis using an RBF classifier for the same combinations of features as presented for the MNN cases; the results are comparable to those of multifeature analysis using the MNN classifier except for one case. Changes in the architecture of the MNN and the RBF classifiers had insignificant effects on the results of multifeature analysis.

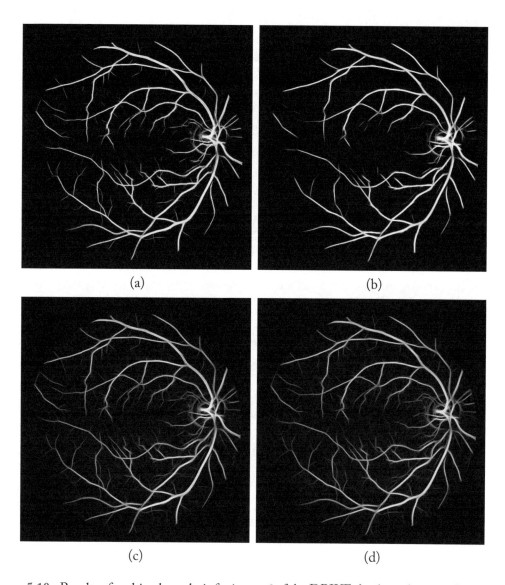

(a) (b)

(c) (d)

Figure 5.10: Results of multiscale analysis for image 6 of the DRIVE database shown as discriminant grayscale images obtained using (a) an MNN with $\tau = \{4, 8, 12\}$, $A_z = 0.9520$; (b) an MNN with $\tau = \{8, 12\}$, $A_z = 0.9473$; (c) an RBF with $\tau = \{4, 8, 12\}$, $A_z = 0.9503$; and (d) an RBF with $\tau = \{0.5, 4, 8, 12\}$, $A_z = 0.9501$. All four cases have approximately the same $A_z \approx 0.95$.

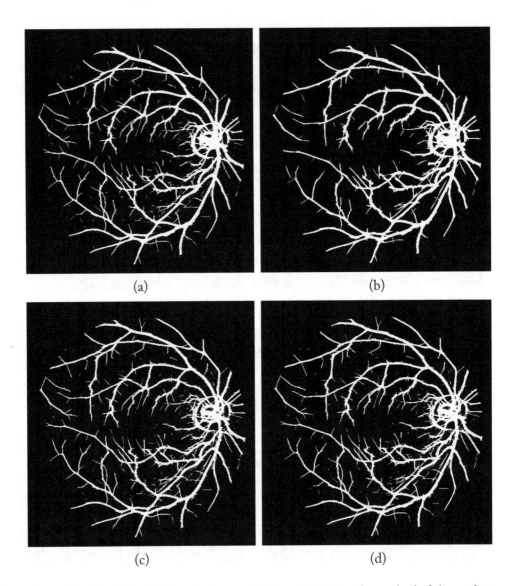

(a)   (b)

(c)   (d)

Figure 5.11: Results of thresholding the images in Figure 5.10 using the method of closest distance to the ROC curve to obtain an optimal threshold for each case: (a) image in Figure 5.10 (a) thresholded at 0.1179; (b) image in Figure 5.10 (b) thresholded at 0.1001; (c) image in Figure 5.10 (c) thresholded at 0.1698; and (d) image in Figure 5.10 (d) thresholded at 0.1779 of the normalized intensity value. In all cases, some pixels belonging to the circular edge related to the ONH have been misclassified as vessels.

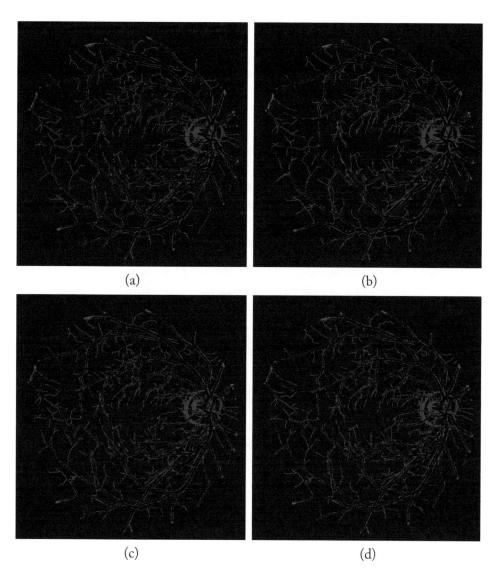

(a)   (b)

(c)   (d)

Figure 5.12: FP (blue) and FN (red) pixels for the results in Figure 5.11, as compared to the ground truth for image 6 of the DRIVE database, as shown in Figure 5.8 (a). In all cases, large numbers of FP pixels (blue) are present around the ONH.

Using the coherence or the inverted G component with a single-scale Gabor magnitude output ($\tau = 8$) did not lead to any improvement, as compared to the results of single-scale analysis. Feature selection techniques (sequential feed-forward selection and stepwise regression [167]) were applied to select appropriate features among seven different Gabor scales, coherence, and the inverted G component; both feature selection methods selected all of the nine features to be included in the training step. However, no improvement was achieved by using all of the available features, as compared to the results presented in the current section.

Table 5.5: Efficiency of the detection of blood vessels (A_z) with multifeature analysis of the test set (20 images) of the DRIVE database using the MNN classifier with different Gabor scale and feature combinations. All MNN classifiers were trained with 15% of the available training data and have one hidden layer with 30 nodes. All hidden nodes have a tansig activation function; the output function for all cases was a linear function. In all instances $l = 2.9$ was used

Features	A_z
$\tau = 8$, Coherence	0.95
$\tau = 8, 1 - G$	0.95
$\tau = \{4,8,12\}$, Coherence	0.96
$\tau = \{4, 8, 12\}, 1 - G$	0.96
$\tau = \{2, 4, 8\}, 1 - G$	0.96
$\tau = \{4,8,12\}$, Coherence, $1 - G$	0.96

Table 5.6: Efficiency of the detection of blood vessels (A_z) with multifeature analysis of the test set (20 images) of the DRIVE database using the RBF classifier with different scale and feature combinations. All RBF classifiers were trained with 0.125% of the available training data and have eight nodes in the hidden layer. In all instances $l = 2.9$ was used

Features	A_z
$\tau = 8$, Coherence	0.95
$\tau = 8, 1 - G$	0.95
$\tau = \{4,8, 12\}$, Coherence	0.96
$\tau = \{4,8, 12\}, 1 - G$	0.96
$\tau = \{2,4, 8\}, 1 - G$	0.95
$\tau = \{4,8, 12\}$, Coherence, and $1 - G$	0.96

Figure 5.13 illustrates examples of discriminant images obtained using multifeature analysis with the MNN and the RBF classifiers. Similar to the multiscale cases, the results of the RBF classifiers [Figure 5.13 (c) and (d)] show lower discrimination between vessel and nonvessel (background) pixels, as compared to results of the MNN classifiers [Figure 5.13 (a) and (b)]. The A_z values in the caption of the figure are related to each individual discriminant image.

Figure 5.14 shows three ROC curves obtained with single-scale, multiscale, and multifeature analysis of the test set of 20 images in the DRIVE database. The multiscale and multifeature analysis procedures have provided comparable performance in the detection of blood vessels; both show marginally better performance, as compared to the results of single-scale analysis.

5.6   COMPARATIVE ANALYSIS

Measures of the efficiency of detection of blood vessels in the retina obtained by several recently reported methods, in terms of A_z values, using the same set of 20 images in the test set of the DRIVE database, are listed in Table 5.7. The results obtained by the methods described in the present work closely match those obtained by Soares et al. [85]. The major differences between the methods described in the present work and those of Soares et al. [85] are: the use of real Gabor filters (instead of complex Gabor filters); the use of a simple MNN that does not assume a Gaussian mixture model; and the use of the inverted luminance component (Y) instead of the G component of the color fundus images. The levels of accuracy obtained in the detection of blood vessels, in terms of A_z values, using the G component are similar to or marginally lower than those obtained using the Y component, as shown in Section 5.3. The use of real Gabor filters takes advantage of the fact that the nature of the contrast of the blood vessels in retinal images is known (that is, positive contrast in the inverted Y component).

As shown in Figure 5.12, most of the errors in the detection of blood vessels in the present work could be attributed to a large number of FP pixels (blue pixels) around the ONH; multiscale analysis and multifeature analysis do not have a significant impact on this source of error. Regardless, as shown in Table 5.7, no other method published to date has exceeded the level of $A_z = 0.96$ obtained in the present work by performing multiscale or multifeature analysis.

Using the directional information provided by the Gabor filters, Dhara et al. [91] applied postprocessing steps to the results of single-scale Gabor filtering to reduce the effect of FP pixels attributed to the boundary of the ONH; after binarization of the Gabor magnitude-response images, the results showed no substantial improvement in terms of sensitivity and specificity.

5.7   REMARKS

The results presented in this chapter represent new and thorough investigations conducted as part of the research work underlying the present book, based on earlier related work reported by Rangayyan et al. [86, 146, 148] and Oloumi et al. [147].

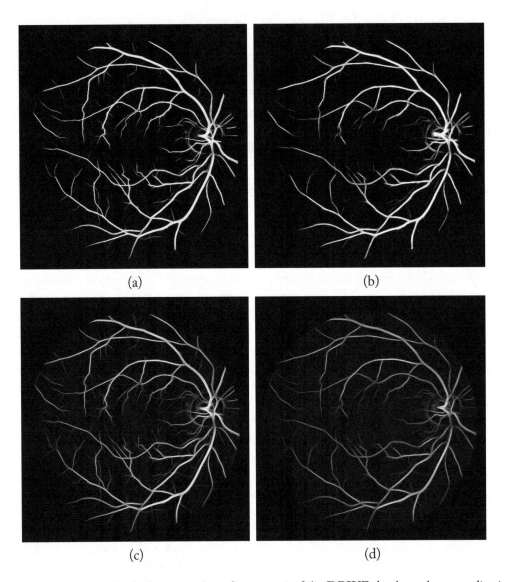

(a)

(b)

(c)

(d)

Figure 5.13: Results of multifeature analysis for image 6 of the DRIVE database shown as discriminant grayscale images generated using (a) an MNN with $\tau = \{4, 8, 12\}$ and the inverted G component, $A_z = 0.9524$; (b) an MNN with $\tau = 8$ and coherence, $A_z = 0.9433$; (c) an RBF with $\tau = \{2, 4, 8\}$ and the inverted G component, $A_z = 0.9474$; and (d) an RBF with $\tau = \{4, 8, 12\}$, coherence, and the inverted G component, $A_z = 0.9523$.

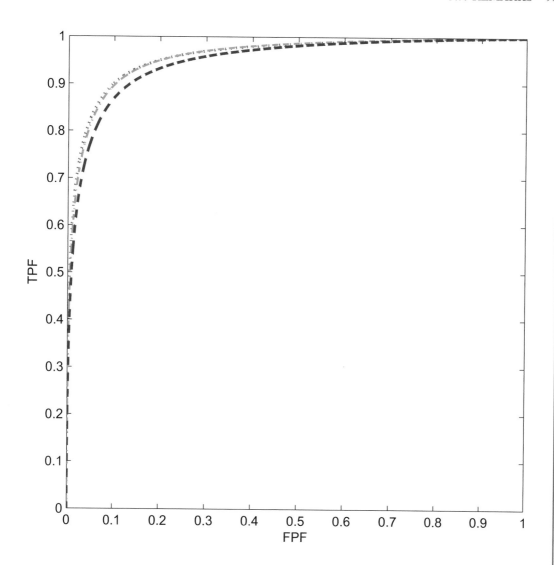

Figure 5.14: Three ROC curves for the test set of 20 images from the DRIVE database obtained using single-scale, multiscale, and multifeature analysis; all of the analyses were performed using the Y component with $l = 2.9$. Dashed blue line: single-scale analysis with $\tau = 8$ resulting in $A_z = 0.9495$. Dot-dashed green line: multiscale analysis with an MNN classifier using $\tau = \{4, 8, 12\}$ having $A_z = 0.9597$. Dotted red line: multifeature analysis with an MNN classifier using $\tau = \{4, 8, 12\}$, coherence, and the inverted G component with $A_z = 0.9606$.

Table 5.7: Comparison of the efficiency of detection of blood vessels in the retina obtained by different methods for the test set (20 images) of the DRIVE database [168]

Detection method	A_z
Matched filter: [71]	0.91
Adaptive local thresholding: [84]	0.93
Ridge-based segmentation: [77]	0.95
Single-scale real Gabor filters using the Y component: [146] and present work	0.95
Single-scale real Gabor filters using the G component: [86] and present work	0.95
Multiscale complex Gabor filters using the G component: [85]	0.96
Multiscale real Gabor filters using the G component: [86]	0.96
Multiscale real Gabor filters using the Y component: [148] and present work	0.96
Multifeature segmentation using 7 features: [89]	0.96
Multifeature segmentation using 41 features: [92]	0.96
Multifeature segmentation using 4 features: present work	0.96

Single-scale Gabor filtering has provided high accuracy in the detection of blood vessels with an area under the ROC curve of $A_z = 0.95$, with the 20 images of the test set of the DRIVE database. Multiscale and multifeature analyses have led to marginal improvements in terms of the area under the ROC curve ($A_z = 0.96$), as compared to the results of single-scale analysis. The differences between the results of multiscale and multifeature analysis are minimal. The advantages to be gained by performing multiscale and multifeature analyses, as compared to single-scale analysis, are also minimal. The results of single-scale analysis as described in the present chapter are used for modeling of the MTA, as explained in Section 6.1.

CHAPTER 6

Modeling of the Major Temporal Arcade

The parabolic or semielliptical profile of the MTA allows for effective modeling using a form of the GHT [118, 185, 186]. In such a model, changes in the TAA could be expected to be reflected as changes in the openness parameter of the parabola: this approach forms the basis for the modeling of the MTA as presented in this chapter [118, 185–187].

All of the methods that have been used to model the MTA, as reviewed in Section 2.1.3, need some type of statistical or geometrical information about the blood vessels. This information can be in the form of manually placed points on the MTA to derive a PDM, or it can be derived automatically from the image using various concepts. As mentioned in Section 2.1.3, two studies have used the GHT to model the MTA automatically: Kochner et al. [82] used extracted edge points to derive the GHT and model the MTA as an ellipse, whereas Fleming et al. [72] used a VST to model the MTA as a semiellipse. A one-pixel-thick skeleton of only the MTA would be the best choice to derive the GHT; thick lines or double edges may result in the detection of multiple curves, as shown in a preliminary study [185]. Using large enough values of the thickness parameter of the Gabor filters (see Section 3.2.2) to emphasize the MTA, it is possible to obtain a skeleton of the VST that contains mostly the MTA, as demonstrated in Section 6.1; such a skeleton would be a suitable choice to derive the GHT.

Based on the anatomical structure of the MTA, certain restrictions could be imposed on the GHT to reduce the computational cost and also to help guide the modeling procedure. The simplest form of implementation of the GHT, without applying any restrictions, is described in Section 6.2. Various anatomical restrictions that could be imposed on the modeling procedure are discussed in Section 6.3.

By using a weighting procedure while incrementing each accumulator cell in the Hough space, it is possible to emphasize the presence of pixels that belong to the MTA, as discussed in Section 6.4.

The global maximum in the Hough space may not always provide the best fit to the MTA; a method to select the appropriate candidate model in the GHT process is described in Section 6.5.

Correction of the retinal raphe angle (see Section 1.1.2) could aid in improving the modeling results, as shown in Section 6.6.

The asymmetric nature of the ITA and the STA implies that separate modeling and analysis of the STA and the ITA could be more beneficial, as demonstrated in Section 6.7. Results of the modeling of the MTA, STA, and ITA are presented in Section 6.8.

6.1 DERIVATION OF THE SKELETON OF THE MTA

The steps required for the acquisition of a skeleton image from the Gabor magnitude-response image are considered to be preprocessing steps to the application of the GHT for the detection of parabolas; an algorithmic flowchart representation of the simplest form of the preprocessing module used in the present work is shown in Figure 6.1. The preprocessing module takes the Gabor magnitude response, as provided by single-scale Gabor filters, as an input. It is demonstrated in Chapter 5 that the design parameters of the real Gabor filters can be fine tuned to detect blood vessels of certain widths. The MTA is the thickest branch of the blood vessels in the retina; therefore, to emphasize its presence and to reduce the influence of thinner vessels, in the present application, single-scale Gabor filters (see Section 5.3) with $\tau = 16$ pixels (0.32 mm) and $l = 2$ are used to detect the MTA. The resulting Gabor magnitude-response image is thresholded by the preprocessing module at 0.0095 of the maximum intensity value to obtain a binary image. Methods to select a threshold automatically, such as Otsu's method [152], for binarization of the Gabor magnitude response led to poorer results, as shown in a related preliminary study [185]; for this reason, a fixed threshold is chosen empirically in the present work. The binarized image is skeletonized using the curvature-skeleton algorithm (see Section 3.1.4). Undesired short segments of 8-connected white pixels having an area less than 70 pixels are removed by using the area-open procedure (see Section 3.1.5). The cleaned and skeletonized representation of the MTA is passed on to the GHT module.

Figure 6.2 illustrates the results of the steps taken by the GHT-preprocessing module. The Gabor magnitude-response image as shown in Figure 6.2 (a) emphasizes the MTA and reduces the presence of thinner vessels; as a result, the thresholded image, as shown in part (b) of the same figure, contains mostly the MTA. Figure 6.2 (c) represents the skeleton (centerline) of the MTA and small segments of other thick vessels. The area-open operation removes some of the small segments, as shown in Figure 6.2 (d).

6.2 PARABOLIC MODELING OF THE MTA

Fleming et al. [72] and Kochner et al. [82] assumed semielliptic and elliptic profiles for the MTA, respectively. However, the MTA diverges away from the ONH toward the macula, and then converges into the macular region. Furthermore, after the second or third branching point, it becomes difficult to distinguish between the original arcade and the new branch as they both become similar in diameter and can branch in unexpected directions. The second and third branching points occur approximately over the macula. If one were to use a parametric curve to fit a model to the entire MTA, it would be difficult to define a specific model; an ellipse appears to be the best esti-

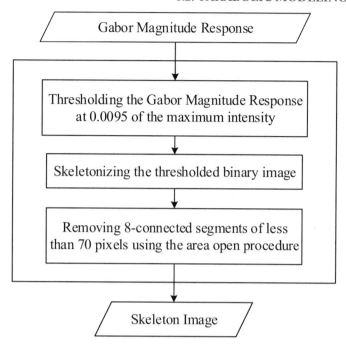

Figure 6.1: Flowchart representation of the GHT-preprocessing module; the preprocessing routine takes the Gabor magnitude-response image as an input and returns the skeleton image of the MTA.

mate to the overall vascular structure, but not a specific arcade. As shown in a related preliminary study, a parabola can effectively model the MTA using the GHT [185].

Figure 6.3 shows a flowchart representation of the GHT module; the input is a skeleton image, such as that shown Figure 6.2 (d), as provided by the GHT-preprocessing module. The GHT module returns the resulting Hough space as the output. In this version of the GHT, as explained in Section 3.3.2, each accumulator cell is incremented by one; hence, this version of the GHT shall be referred to as the unity-updated GHT procedure. As mentioned in Section 3.3, the Hough space is constructed based on the parameters that govern the formula of the parametric curve to be detected; in the case of a parabolic form, the parameters are (m_o, n_o, a). In the present work, the limits of the (m_o, n_o) parameters are set equal to the height and width (rows and columns) of the skeleton image that is used to derive the GHT. As mentioned in Section 3.3.2, the parameter a is, in theory, unbounded. However, in the present work, the value of a is restricted by physiological limits on the MTA and the size of the image. For the DRIVE database, $|a|$ is confined to the range [35, 120]. In order not to make the accumulator too large, only positive values of a are permitted. If an image has the MTA opening to the left, it is rotated by 180° so

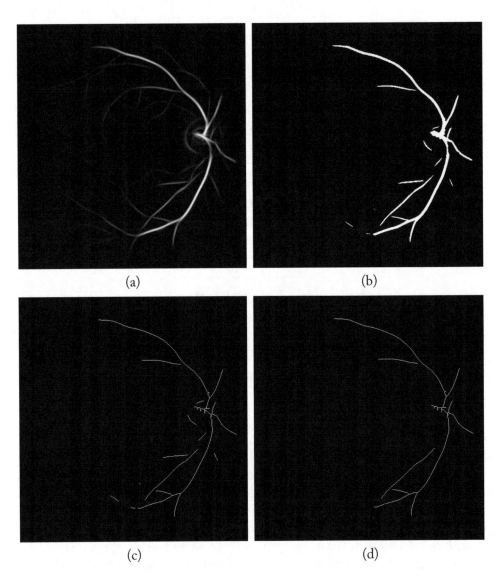

(a)   (b)

(c)   (d)

Figure 6.2: (a) Gabor magnitude response for the image in Figure 5.5 (c) obtained using $\tau = 16$ and $l = 2$. (b) A binarized representation of the MTA obtained by thresholding the image in (a) at 0.0095 of the maximum intensity. (c) Result of skeletonizing the binary image in (b). (d) Application of the area-open procedure has removed small unwanted segments in (c) that have an area less than 70 pixels.

that the MTA opens to the right in the image used for the subsequent steps. Thus, the number of planes in the Hough space for the parameter a is defined to be $120 - 34 = 86$. Therefore, for the simplest form of implementation of the GHT, as described in the current section, the size of the Hough space accumulator is set to be $(584, 565, 86)$, where $(584, 565)$ is the size of the skeleton image used to derive the GHT, as shown in Figure 6.2 (d).

The process of finding the appropriate parabolic-model candidate from the Hough space is considered to be a postprocessing step; Figure 6.4 shows a flowchart representation of the postprocessing module for selection of the index of the global maximum in the Hough space, (m_o, n_o, a), as the parameters of the best-fitting parabolic model.

Figure 6.5 shows the result of applying the simplest form of the unity-updated GHT for parabolic modeling, as explained in Section 3.3.2, and performed in a preliminary study [185], using the skeleton image in Figure 6.2 (d), as provided by the GHT-preprocessing module. Figure 6.5 (a) shows the Hough-space plane for $a = -42$, where the global maximum resides; the index of the point with the highest value in this Hough-space plane, $(295, 563)$, indicates the location of the vertex of the detected parabola, and the value of the Hough-space plane, $a = -42$, indicates the value of the openness parameter a. Note that the top-left corner of the image is represented by the point$(1, 1)$. The display intensity for all the Hough-space figures in the present chapter are $\log_{10}(1 + \text{accumulator cell value})$. Figure 6.5 (b) shows the resulting parabolic model, in green, overlaid on the original image 6 of the DRIVE database. It can be observed that the detected vertex falls outside the FOV because no restriction has been imposed on the vertex in the unity-updated GHT algorithm; as a result, the parabolic model is only matching the outer parts of the MTA close to the macular region.

6.3   ANATOMICAL RESTRICTIONS ON THE GHT

As shown in a preliminary study [185], the use of the GHT for the detection of parabolas in retinal images without imposing restrictions on the GHT procedure could have unwanted consequences, such as the vertex of the parabolic model falling outside of the FOV, as shown in Figure 6.5 (b). Such drawbacks indicate the need for certain restrictions on the GHT process for more accurate modeling.

There are three different anatomical restrictions placed on the GHT process in the present work. The first restriction, which is used to set upper and lower bounds on the value of the a parameter (see Section 6.2), is essential in order for the GHT process to work, in general. However, the other two restrictions, as discussed in the current section, are used to assist the GHT modeling algorithm, as well as to reduce the computational cost of the GHT algorithm.

6.3.1   HORIZONTAL RESTRICTION

The fact that the posterior changes that occur to the MTA are closer to the ONH, along with a priori knowledge of the center of the ONH, and the observation that the macula is situated approximately two ONHDs temporal to the ONH [19], are used in the present work to guide

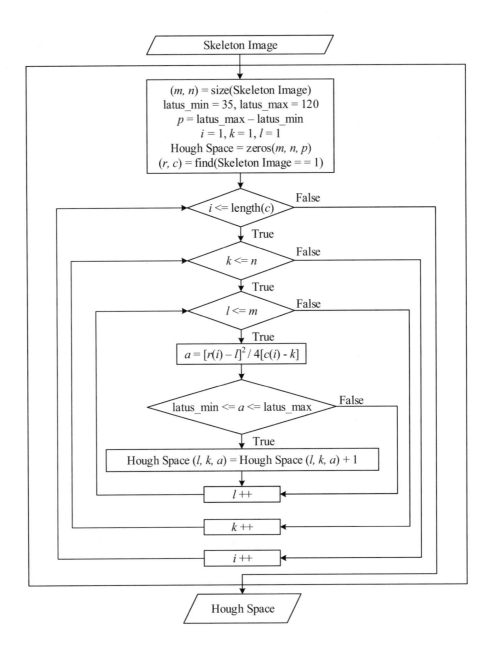

Figure 6.3: Flowchart representation of the unity-updated GHT module.

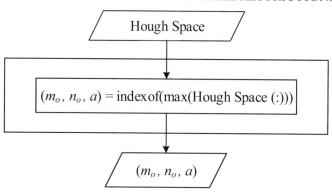

Figure 6.4: A flowchart representation of the GHT-postprocessing module; the postprocessing routine finds the indices of the point in the Hough space that has the highest value and returns them as the parameters of the detected parabolic model.

the modeling procedure. The average ONHD in adults is about 1.6 mm [19, 20], which is approximately 80 pixels for the images of the DRIVE database. The MTA has a parabolic shape up to the macula; therefore, for parabolic modeling, a skeleton of the MTA is required only from the ONH to the macular region for parabolic modeling. There appears to be no useful information on the nasal side of the ONH; thus, this part may also be eliminated. Hence, in the present work, the Gabor magnitude image was horizontally limited from 0.25×ONHD nasal to 2×ONHD temporal with respect to the center of the ONH. Using the average ONHD and given the spatial resolution of the DRIVE database [168], the image was horizontally limited to the range $[ONH_n - 20, ONH_n + 160]$, or (584, 181) pixels, where ONH_n denotes the column-coordinate of the center of the ONH, detected automatically using Gabor filters and phase portrait analysis, as described by Rangayyan et al. [116]. The cropped skeleton image is obtained by applying the Gabor magnitude-response image to the GHT-preprocessing module with horizontal cropping, as shown in Figure 6.6. By cropping the skeleton image horizontally, the Hough-space dimensions are changed to (584, 181, 86), which significantly reduces the computational cost of the GHT algorithm.

Figure 6.7 shows the result of applying the unity-updated GHT with horizontal restriction to image 6 of the DRIVE database. Figure 6.7 (a) shows the cropped Gabor magnitude response; the parabolic form of the MTA is emphasized in this manner. Part (b) of the same figure shows the cleaned skeleton of part (a), as obtained by the preprocessing module. The cropped skeleton image is applied to the unity-updated GHT algorithm, as shown in Figure 6.3. Figure 6.7 (c) shows the Hough-space plane for $a = -70$, where the global maximum resides; the vertex of the detected parabolic model is given by the coordinates (271, 460). Figure 6.5 (d) shows the resulting

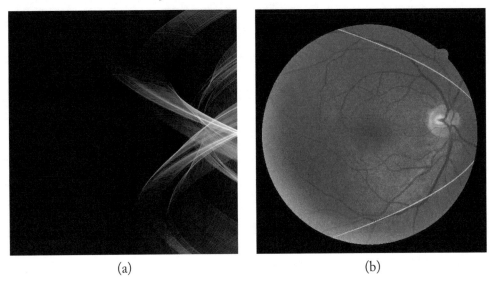

(a)   (b)

Figure 6.5: Illustration of the result of applying the simplest form of the unity-updated GHT to image 6 of the DRIVE database: (a) The Hough-space plane for $a = -42$ containing the global maximum in the Hough space. The index of the point with the highest value, along with the value of the a parameter, make up the parameters of the detected parabolic model. (b) The parabolic model obtained from (a), specified by the parameters $(295, 563, -42)$, is shown overlaid on the original image 6 of the DRIVE database; the detected vertex of the parabolic model falls outside of the FOV because no restrictions are applied on the vertex in the unity-updated GHT algorithm.

parabolic model, given by the parameters $(271, 460, -70)$, overlaid on the original image 6 of the DRIVE database; the detected model is more accurate than the model in Figure 6.5, even though the parabolic model is only matching the STA. On closer inspection, it becomes clear that the model is matching the inferior arteriole, which has not been eliminated by either the thresholding process or the area-open procedure.

Further restrictions and modifications that are applied to the GHT process, as explained in rest of the current chapter, are only applied to the horizontally restricted version of the GHT algorithm.

6.3.2   VERTEX RESTRICTION

To make further use of the possible anatomical restrictions, the vertex location, (m_o, n_o), of the parabolas represented in the Hough space was restricted to be within a disk of radius $r = 0.25 \times \text{ONHD}$ centered at the previously detected center of the ONH, as described by Rangayyan et al. [116]. The accumulator cells in the Hough space are incremented only if the indices

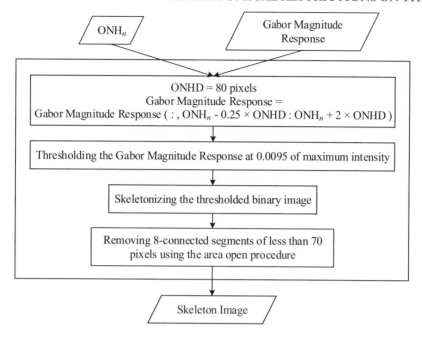

Figure 6.6: A flowchart representation of the GHT-preprocessing module with horizontal cropping. The preprocessing routine takes the Gabor magnitude image as an input and returns the horizontally cropped skeleton of the MTA.

of the corresponding vertex fall within a disk of radius 20 pixels with its center at the detected center of the ONH. Figure 6.8 shows a flowchart representation of the module responsible for the generation of a mask image to indicate where the vertex is allowed to fall; the procedure takes the location of the detected center of the ONH and the FOV mask created by the preprocessing module of the Gabor procedure, as inputs. The FOV mask is needed in case the disk of radius $r = 0.25 \times \text{ONHD}$ has parts that fall outside of the FOV.

The inputs to the GHT module with vertex restriction are the skeleton image, as provided by the modified preprocessing module (see Figure 6.6), and the Vertex Mask image indicating the area in which the vertex is allowed to fall. The flowchart representation of the GHT module with vertex restriction is similar to the flowchart shown in Figure 6.3, except that, before calculating the openness parameter, a, the algorithm checks to see if Vertex Mask(l, k) $== 1$; if the statement is true, the algorithm carries on with the calculation of the openness parameter; if the statement is false, the current iteration is skipped. The vertex restriction as applied in the present work further reduces the computational cost of the GHT procedure.

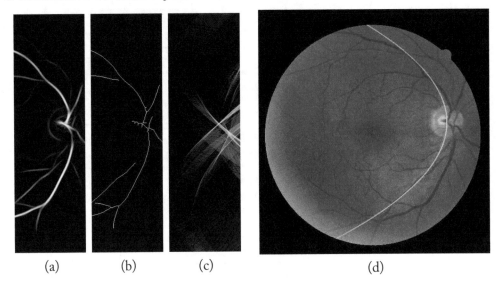

(a)   (b)   (c)   (d)

Figure 6.7: Results of applying the horizontally limited and unity-updated GHT to image 6 of the DRIVE database. (a) Result of horizontally cropping the Gabor magnitude response. (b) The skeleton image, provided by the modified preprocessing module (Figure 6.6), is clipped horizontally to the range $[ONH_n - 0.25 \times ONHD, ONH_n + 2 \times ONHD]$. (c) Hough-space plane for $a = -70$; the vertex of the best-fitting parabola, as selected using the postprocessing module, shown in Figure 6.4, is at coordinates $(271, 460)$ in the original full-size image. The parabolic model is given by the parameters $(271, 460, -70)$. (d) Parabolic model given in (c) drawn in green over the original color image; the inferior part of the model is matching the inferior arteriole instead of the ITA.

Figure 6.9 shows the results of applying the unity-updated GHT with vertex restriction to image 6 of the DRIVE database. The horizontally cropped vertex-restricting mask, as provided by the vertex mask generator module shown in Figure 6.8, is illustrated in Figure 6.9 (b); the vertices of the potential parabolas in the Hough space are required to be located within the white disk. The Hough-space plane containing the point with the highest value is shown in Figure 6.9 (c). The resulting parabolic model is shown in Figure 6.9 (d). The parabolic model is matching the inferior arteriole instead of the ITA.

6.4   WEIGHTING OPTIONS FOR THE GHT

Incrementing each accumulator cell value by one in the unity-updated GHT procedure implies giving equal vote to all of the pixels in the skeleton image. However, the skeleton image is not made up of only the MTA. It is possible to give more weight to the votes of pixels that belong to the MTA by using the Gabor magnitude-response image; the higher the Gabor magnitude response

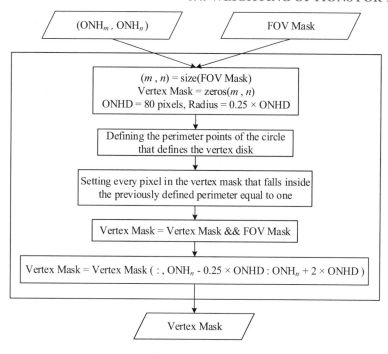

Figure 6.8: A flowchart representation of the module responsible for the generation of the vertex mask, which is used in the vertex-restricted GHT module.

of a pixel, the more likely it is that the pixel belongs to the MTA (given the Gabor parameters used for the present application). Thus, in a variation of the GHT process, instead of incrementing each accumulator cell by unity, it was incremented by the Gabor magnitude response of the same pixel as that indicated by the skeleton image. The Gabor-magnitude-updated GHT routine requires the Gabor magnitude-response image and the skeleton image as inputs and returns the resulting Hough space. The flowchart representation of the Gabor-magnitude-updated GHT module is similar to the unity-updated GHT module (see Figure 6.3), except that instead of incrementing the Hough space cell by 1, it is incremented by the value of the Gabor magnitude response at the corresponding pixel under consideration, i.e., Gabor Magnitude Response$[r(i), c(i)]$.

Figure 6.10 shows the results of applying the Gabor-magnitude-updated GHT module to image 6 of the DRIVE database. Figure 6.10 (c) shows the plane that contains the point with the highest value in the Hough space. Part (d) of the same figure shows the selected parabolic model drawn on the original color image. The parabolic model is fitting the STA closely; however, the inferior part of the model is matching the inferior arteriole and not the ITA. The use of the Gabor magnitude response to update the Hough-space cells increases the accumulator value for

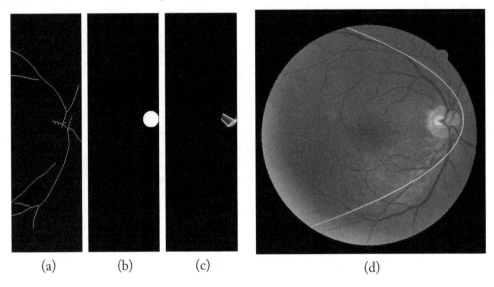

(a)   (b)   (c)   (d)

Figure 6.9: Results of applying the unity-updated GHT with vertex restriction to image 6 of the DRIVE database. (a) Horizontally cropped VST of the MTA. (b) The vertex mask created by the mask generation module shown in Figure 6.8. (c) The Hough-space plane for $a = -43$ containing the point with the highest value in the Hough space given by the indices $(265, 509)$. (d) The selected parabolic model, with the parameters $(265, 509, -43)$, drawn in green over the original color image; the parabolic model is matching the inferior arteriole instead of the ITA.

parabolas originating from the thickest branch. Hence, it produces a less-sparse Hough space with bright spots only close to the ONH.

6.5   SELECTION AMONG THE HOUGH-SPACE CANDIDATES

The postprocessing routine, as described in Section 6.2, selects the point with the highest value in the resulting Hough space to obtain the parameters (m_o, n_o, a) of the best-fitting parabolic model to the MTA. However, the global maximum in the Hough space may not always present the best-fitting model. Hence, a procedure based on the MDCP measure (see Section 4.3.2) is implemented to select the best fit among the top 10 Hough-space candidates. The MDCP values are calculated for each of the top 10 parabolic fits with the automatically detected VST serving as the reference. The fit with the smallest MDCP value is selected as the best-fitting model to the MTA. Figure 6.11 shows a flowchart representation of the postprocessing module based on the MDCP measure.

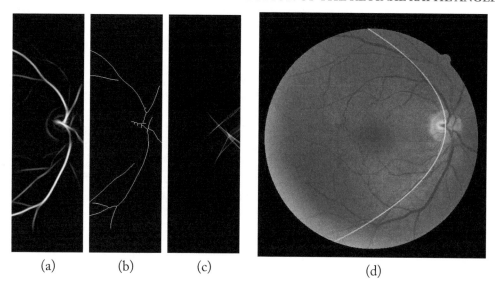

(a)   (b)   (c)   (d)

Figure 6.10: Results of applying the Gabor-magnitude-updated GHT to image 6 of the DRIVE database. (a) The horizontally cropped Gabor magnitude-response image used to update the Hough-space accumulator cells. (b) Horizontally cropped VST of the MTA. (c) The Hough-space plane for $a = -70$ containing the point with the highest value in the Hough space, given by the indices $(271, 460)$. (d) The selected parabolic model, with the parameters $(271, 460, -70)$, drawn in green over the original color image; the parabolic model is matching the inferior arteriole instead of the ITA. The parameters of the parabolic model obtained using the Gabor-magnitude-updated GHT procedure are the same as those obtained using the unity-updated GHT procedure, as shown in Figure 6.7.

Figure 6.12 shows the result of applying the MDCP-based postprocessing procedure to the unity-updated GHT algorithm for image 6 of the DRIVE database. Figure 6.12 (b) shows the Hough-space plane for $a = -50$ for the selected model based on the lowest-MDCP-value criterion. The resulting parabolic model, given by the parameters $(268, 492, -50)$, is the sixth model among the top 10 Hough-space candidates; the selected parabolic model is shown in Figure 6.12 (c). The parabolic model is fitting the STA, but not the ITA.

6.6   CORRECTION OF THE RETINAL RAPHE ANGLE

Any rotation that might exist between the retinal raphe and the horizontal axis of the given image could affect the modeling procedure. The manually marked fovea and the center of the ONH are used in the present work to determine the raphe angle and correct the image by the same amount. Calculation of the raphe angle is illustrated and described in Figure 1.6. The raphe-angle-correction step is applied to the original color images of the retina before the application of

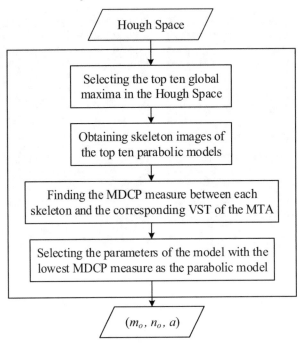

Figure 6.11: A flowchart representation of the GHT-postprocessing module performing the MDCP-based selection method. The postprocessing routine finds the indices of the top 10 points in the Hough space that have the highest values. Skeleton images of the top 10 parabolic models are obtained and then the MDCP measures are computed from each of the parabolic models to the corresponding VST of the MTA. The model that results in the lowest MDCP measure is selected as the most accurate parabolic model.

the Gabor-preprocessing and the Gabor filtering modules; the resulting mask, Gabor magnitude response, and skeleton images will all be raphe-angle-corrected in this manner.

Figure 6.13 shows the result of applying the unity-updated GHT procedure with the raphe-angle-correction step to image 6 of the DRIVE database. Figure 6.13 (b) shows the Hough-space plane containing the point with the highest value in the Hough space. The resulting parabolic model is shown in part (c) of the same figure; the superior part of the model is not closely fitting the STA and the inferior part is not modeling the ITA.

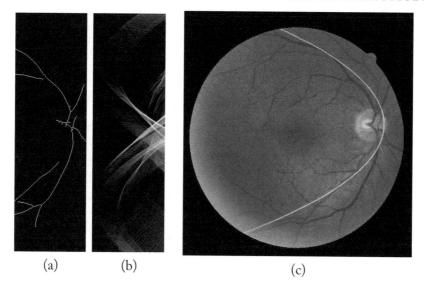

$$(a) \qquad (b) \qquad\qquad (c)$$

Figure 6.12: Results of applying the unity-updated GHT routine with the MDCP-based selection option to image 6 of the DRIVE database. (a) Horizontally cropped VST of the MTA. (b) Hough-space plane for $a = -50$, containing the selected vertex of the best-fitting parabolic model at the location $(268, 492)$, obtained using the MDCP-based selection option. (c) The resulting parabolic model, given by the parameters $(268, 492, -50)$, shown in green over the original color image. The parabolic model is matching the inferior arteriole instead of the ITA.

6.7   DUAL-PARABOLIC MODELING

The ITA and the STA are often asymmetric; thus, a single-parabolic model may match either one of the arcades, but not both. Modeling each part of the arcade separately may be a more suitable option. The GHT is a versatile method, and in the absence of full parabolas in an image, the method may be used to estimate the vertices and the openness parameters of semiparabolas. For the purpose of dual-parabolic modeling, the automatically detected center of the ONH is used to separate the Gabor magnitude-response image into its superior and inferior parts. Figure 6.14 shows a flowchart representation of the preprocessing module used in the present work to obtain separate skeleton images of the STA and ITA. To represent the ITA, any information in the range $m \in [1, \text{ONH}_m]$ is eliminated in the Gabor magnitude-response image. Any information within the range $m \in [\text{ONH}_m + 1, 584]$ is eliminated in the Gabor magnitude-response image to obtain a representation of the STA. The same procedures as in the preprocessing module in Section 6.3 are applied to the superior and inferior Gabor magnitude-response images to obtain the horizontally limited VSTs of the STA and the ITA, respectively.

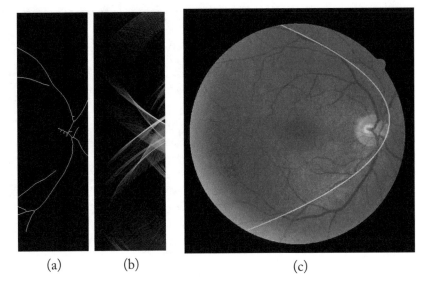

(a)   (b)   (c)

Figure 6.13: Results of applying the unity-updated GHT routine with raphe-angle correction to image 6 of the DRIVE database. (a) Horizontally cropped VST of the MTA. (b) The Hough-space plane for $a = -49$ containing the point with the highest value in the Hough space. (c) The parabolic model given by the parameters $(268, 503, -49)$ is drawn on top of the original color image 6.

When modeling each arcade separately, it is essential to restrict the vertices of the resulting parabolas in the Hough space. Otherwise, the result may only match a straight segment of the MTA. For this reason, only the Gabor-magnitude-updated GHT with the vertex restriction is applied to the skeleton images of the STA and the ITA. Accumulator cells are updated with the Gabor-magnitude response of the same pixel as that indicated by the skeleton image, only if they fall within a disk of radius $r = 0.25 \times$ONHD, centered at the detected center of the ONH.

The Gabor-magnitude-updated GHT with vertex restriction along with raphe-angle correction and the MDCP-based selection procedures were applied to the skeleton images obtained of the STA and the ITA. The part of the fit to the STA in the range $m \in [1, V_m]$ is taken as the STA model, where V_m is the row-coordinate of the detected vertex of the parabola. The ITA model is taken as the fit to the ITA in the range $m \in [V_m, 584]$. This form of modeling results in two openness parameters: a_{STA} and a_{ITA}. In such a dual model, changes in the SAA and the IAA could be expected to be reflected as changes in the openness parameters of the semiparabolas. Separate analysis of the changes in a_{STA} and a_{ITA} may be more beneficial, as shown by the correlation of the stages of ROP with only the changes in the IAA in the work of Wilson et al. [40].

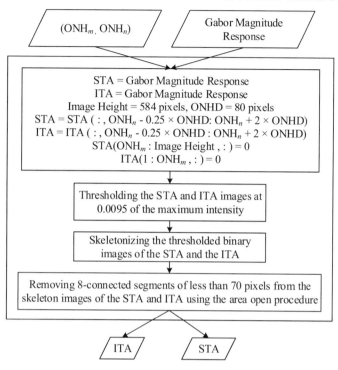

Figure 6.14: A flowchart representation of the preprocessing module for the dual-parabolic modeling procedure. The Gabor magnitude-response image is separated into its superior and inferior parts using the previously detected center of the ONH.

Figure 6.15 shows the results of applying the Gabor-magnitude-updated GHT with vertex restriction for dual-parabolic modeling to image 6 of the DRIVE database. Each of the STA and the ITA skeleton images [Figure 6.15 (a) and (b), respectively], as provided by the preprocessing module shown in Figure 6.14, are provided separately to the Gabor-magnitude-updated GHT module with vertex restriction to obtain two models. Figure 6.15 (c) shows the Hough-space plane that contains the global maximum in the Hough space in the case of the STA. Similarly, part (d) of the same figure shows the Hough-space plane that contains the global maximum in the Hough space in the case of the ITA. The dual-parabolic models selected as the best-fitting models for the STA and the ITA, drawn over the original color image, are shown in Figures 6.15 (e) and (f), respectively. The ITA model is fitting the ITA closely; the STA model is missing a small posterior part of the STA.

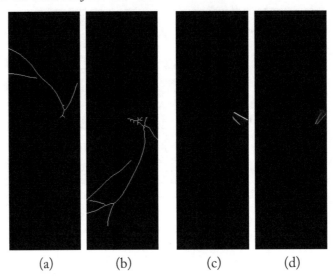

(a)   (b)   (c)   (d)

Figure 6.15: Results of dual-parabolic modeling using the Gabor-magnitude-updated GHT with vertex restriction as applied to image 6 of the DRIVE database. Horizontally cropped VST images of (a) the STA and (b) the ITA, as provided by the preprocessing module shown in Figure 6.14. (c) The Hough-space plane for $a = -44$ containing the point with the highest value in the Hough space for the STA model, given by the indices $(261, 498)$. (d) The Hough-space plane for $a = -108$ containing the point with the highest value in the Hough space for the ITA model, given by the indices $(278, 485)$. *Continues.*

There are four different variations of the GHT module as described in the present work:

- unity-updated GHT (Section 6.2),

- unity-updated GHT with vertex restriction (Section 6.3.2),

- Gabor-magnitude-updated GHT (Section 6.4), and

- Gabor-magnitude-updated GHT with vertex restriction (Section 6.7).

Single-parabolic modeling and testing is performed with all of the versions of the GHT mentioned above, along with the horizontally cropped images, as explained in Section 6.3. For the purpose of dual-parabolic modeling and testing, only the Gabor-magnitude-updated GHT with vertex restriction is used. In addition, the raphe-angle correction and the MDCP-based selection options (separately and combined) are also used for single- and dual-parabolic modeling and testing in the present work; the results are presented in Section 6.8.

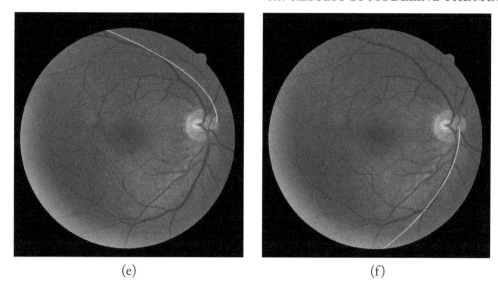

(e)   (f)

Figure 6.15: *Continued.* (e) The selected dual-parabolic model for the STA, with the parameters $(261, 498, -44)$, drawn in green over the original color image; the parabolic model is missing a small posterior part of the STA. (f) The selected dual-parabolic model for the ITA, with the parameters $(278, 485, -108)$, drawn in green over the original color image; the parabolic model is closely fitting the ITA.

6.8   RESULTS OF MODELING THE MTA

6.8.1   ASSESSMENT OF THE ACCURACY OF THE OBTAINED VSTs

As mentioned in Section 4.3, to assess the accuracy of the obtained VSTs, the correlation coefficients of the parameter a and the vertex errors of the parabolic fits for the results of the four versions of the GHT mentioned in Section 6.7, as compared to the parameters of the fits to the corresponding hand-drawn arcades using the unity-updated GHT with vertex restriction, are used in the present work.

Figure 6.16 (a) shows the horizontally cropped hand-drawn trace of the MTA (see Section 4.2) used to obtain a parabolic model via the unity-updated GHT with vertex restriction. Part (b) of the same figure shows the resulting Hough-space plane, which contains the global maximum for the unity-updated GHT with vertex restriction. The resulting parabolic model is shown in Figure 6.16 (c).

The correlation coefficients of the parameter a and the vertex errors of the parabolic fits mentioned above, as compared to the parameters of the fits to the corresponding hand-drawn arcades using the unity-updated GHT with vertex restriction, are given in Table 6.1. The Gabor-magnitude-updated GHT with the vertex restriction led to the highest correlation of the openness

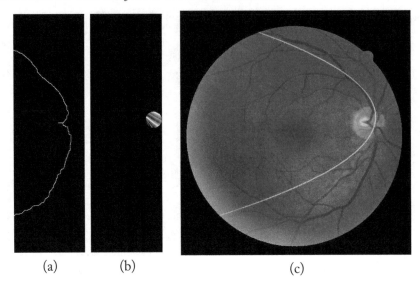

(a)   (b)   (c)

Figure 6.16: Results of applying the unity-updated GHT with vertex restriction to the hand-drawn trace of the MTA for image 6 of the DRIVE database. (a) Horizontally cropped hand-drawn trace of the MTA. (b) The Hough-space plane for $a = -43$ containing the global maximum given by the indices $(265, 509)$. (c) The resulting parabolic model, given by the parameters $(265, 509, -43)$, shown in green over the original color image. The model is matching only a part of the STA.

parameter a, whereas the unity-updated GHT with the vertex restriction provided the lowest error in the detected vertices. These measures are used in the present work only to assess the performance of the procedure that automatically obtains the VST.

6.8.2   ASSESSMENT OF THE ACCURACY OF THE PARABOLIC MODELS

As discussed in Section 4.3, the MDCP and Hausdorff measures are used in the present work to assess the accuracy of the parabolic models obtained using the GHT. The MDCP measure can be illustrated by drawing the DCPs between the model and the reference. However, the Hausdorff measure is not illustrative, as it only represents a single distance measure (largest possible error); hence, only the MDCP error measures are illustrated in the current section. In the following illustrations in Figures 6.17 to 6.22, the red contour represents the parabolic model and the green trace represents the hand-drawn arcade for image 6 of the DRIVE database; the only exception is Figure 6.22, in which image 19 of the DRIVE database is shown. The magenta '*' mark indicates the automatically detected center of the ONH. The yellow vertical line shows the horizontal extent of MDCP measurement from the ONH. The cyan lines connecting the model to the hand-drawn

Table 6.1: The correlation coefficients and the average vertex errors between the parameters of the parabolas obtained with the four different GHT versions, compared to the parameters of the fits obtained using the unity-updated GHT applied to the hand-drawn arcades for the 40 DRIVE images. The average vertex errors and their STDs are provided in terms of pixels, where each pixel is 20 μm. Note that VR stands for vertex restriction and GM stands for Gabor-magnitude

GHT Version	Correlation Coefficient	Vertex Error, Mean ± STD (Pixels)
Unity-updated	0.91	37.65 ± 35.17
Unity-updated with VR	0.96	10.61 ± 8.18
GM-updated	0.92	36.82 ± 25.18
GM-updated with VR	0.97	12.53 ± 9.2

trace of the arcade are the DCPs. The DCP is drawn in the illustrations only for every fifth point on the model.

Figure 6.17 illustrates the DCP errors for the results of different versions of the unity-updated GHT, with and without the MDCP-based selection, vertex restriction, and raphe-angle correction options, as applied to image 6 of the DRIVE database. The lower MDCP error for the result in Figure 6.17 (a), as compared to the results in parts (b), (c), and (d), is apparent by visual comparison.

Figure 6.18 illustrates the DCP errors for the results of the Gabor-magnitude-updated GHT without vertex restriction for image 6 of the DRIVE database, along with, and without raphe-angle correction and MDCP-based selection options. The low MDCP error for the result in Figure 6.18 (b), as compared to the results in parts (a), (c), and (d), is clear by visual inspection.

Figure 6.19 illustrates the DCP errors for the results of the Gabor-magnitude-updated GHT with vertex restriction for image 6 of the DRIVE database, along with, and without MDCP-based selection and raphe-angle correction options. The low MDCP error for the result in Figure 6.19 (d), as compared to the results in parts (a), (b), and (c), is apparent by visual inspection.

In the case of the dual-parabolic modeling procedure, the MDCP is measured from the STA and the ITA models to the hand-drawn traces of the STA and the ITA, respectively; hence, two separate error measures are obtained for the results of the dual-parabolic modeling procedure. For the purpose of dual-parabolic modeling, only the Gabor-magnitude-updated and vertex-restricted GHT is used. Figures 6.20 (a) and (b) illustrate the DCP errors for the STA and the ITA models, obtained without using raphe-angle correction and MDCP-based selection, as compared to the hand-drawn traces of the STA and the ITA, respectively. Similarly, Figures 6.20 (c) and (d)

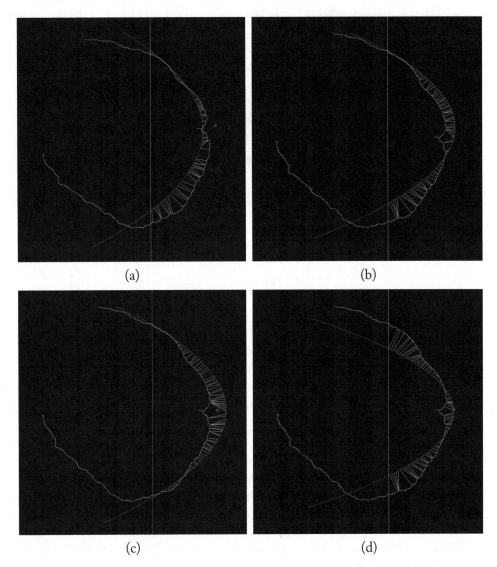

(a)

(b)

(c)

(d)

Figure 6.17: Illustration of the DCP errors for the results of modeling of the MTA for image 6 of the DRIVE database, using various versions of the unity-updated GHT. (a) Model obtained without using any option, resulting in an MDCP error of 16 pixels. (b) Model obtained with raphe-angle correction and MDCP-based selection, resulting in an MDCP error of 19 pixels. (c) Model obtained with vertex restriction and raphe-angle correction, resulting in an MDCP error of 18 pixels. (d) Model obtained with vertex restriction, raphe-angle correction, and MDCP-based selection, resulting in an MDCP error of 22 pixels.

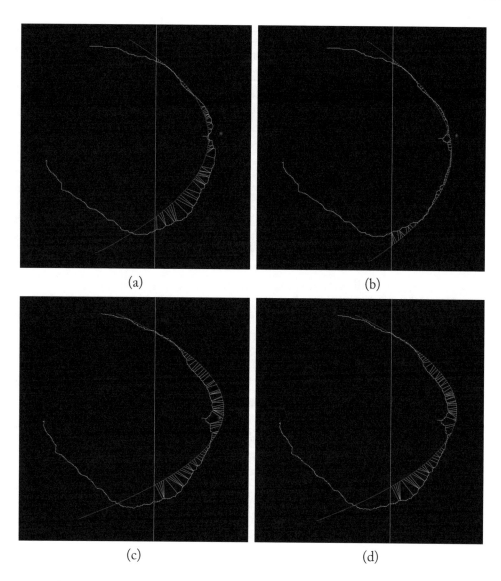

(a)   (b)

(c)   (d)

Figure 6.18: Illustration of the DCP errors for the results of modeling of the MTA for image 6 of the DRIVE database, using Gabor-magnitude-updated GHT without vertex restriction. (a) Model obtained without using any option, resulting in an MDCP error of 16 pixels. (b) Model obtained using MDCP-based selection, resulting in an MDCP error of 5 pixels. (c) Model obtained using raphe-angle correction, resulting in an MDCP error of 19 pixels. (d) Model obtained using raphe-angle correction and MDCP-based selection combined, resulting in an MDCP error of 18 pixels.

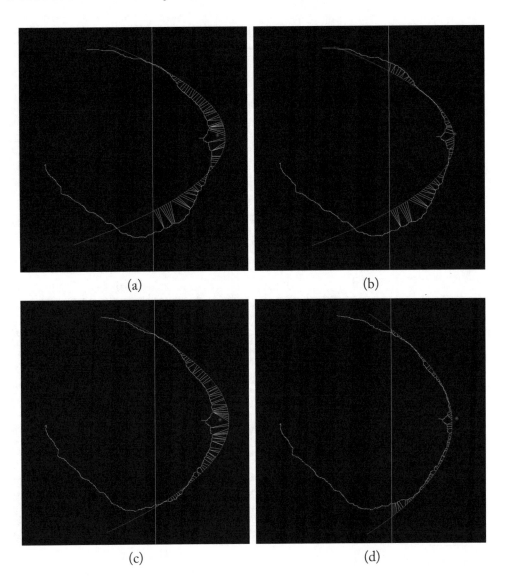

(a)   (b)

(c)   (d)

Figure 6.19: Illustration of the DCP errors for the results of parabolic modeling of the MTA for image 6 of the DRIVE database, using Gabor-magnitude-updated and vertex-restricted GHT. (a) Model obtained without using any option, resulting in an MDCP error of 24 pixels. (b) Model obtained using MDCP-based selection, resulting in an MDCP error of 17 pixels. (c) Model obtained using raphe-angle correction, resulting in an MDCP error of 18 pixels. (d) Model obtained using raphe-angle correction and MDCP-based selection combined, resulting in an MDCP error of 6 pixels.

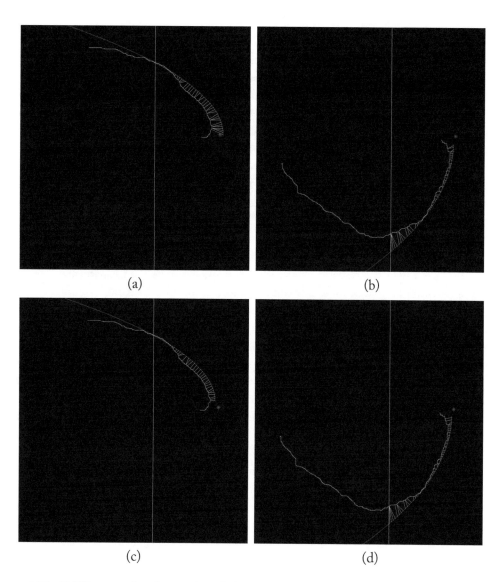

(a)   (b)

(c)   (d)

Figure 6.20: DCP errors for the results of dual-parabolic modeling for image 6 of the DRIVE database, using the Gabor-magnitude-updated and vertex-restricted GHT for (a) the STA and (b) the ITA models, resulting in MDCP errors of 14 and 10 pixels, respectively. (c) The STA and (d) the ITA models obtained using the Gabor-magnitude-updated and vertex-restricted GHT with MDCP-based selection, resulting in MDCP errors of 10 and 9 pixels, respectively.

show the DCP errors for the STA and the ITA models, obtained with MDCP-based selection, as compared to the hand-drawn traces of the STA and the ITA, respectively.

Figures 6.21 (a) and (b) illustrate the DCP errors for the STA and the ITA models, obtained using the Gabor-magnitude-updated and vertex-restricted GHT with raphe-angle correction, as compared to the hand-drawn traces of the STA and the ITA, respectively. Similarly, Figures 6.21 (c) and (d) shows the DCP errors for the STA and the ITA models, obtained using the Gabor-magnitude-updated and vertex-restricted GHT with MDCP-based selection and raphe-angle correction combined, as compared to the hand-drawn traces of the STA and the ITA, respectively. The dual-parabolic model obtained using the Gabor-magnitude-updated and vertex-restricted GHT with MDCP-based selection [Figures 6.20 (c) and (d)] provides the most accurate dual-parabolic model for image 6 of the DRIVE database with an average MDCP error of 9.5 pixels. The single-parabolic model acquired using the Gabor-magnitude-updated GHT with MDCP-based selection has the lowest MDCP error (5.0 pixels) among the single-parabolic models for image 6 of the DRIVE database.

Figure 6.22 demonstrates the DCP errors for the results of dual-parabolic modeling for image 19 of the DRIVE database; for the sake of comparison of the ITA and the STA fits, both models are combined in one image. It can be observed that the STA has a semiparabolic shape, whereas the ITA resembles an exponential function; this point is reinforced by the low MDCP error for the STA model, as compared to the well-above-average MDCP error for the ITA model.

The average MDCP and Hausdorff errors of the parabolic fits for the results of the four different versions of the GHT and the dual-parabolic modeling procedure, as compared to the hand-drawn arcades for all of the 40 DRIVE images, are listed in Tables 6.2 and 6.3, respectively. Both of the error measures also are used to assess the performance of the modeling methods with the added options of raphe-angle correction and MDCP-based selection, separately and combined together. The dual-parabolic modeling procedure provides separate models for the ITA and the STA; hence, the MDCP and the Hausdorff error measures are obtained for each of the ITA and STA models separately, as compared to their corresponding hand-drawn traces. Combining the ITA and the STA models to obtain the MDCP and the Hausdorff errors, for the sake of comparison with the results of the other GHT versions, is not appropriate, as it may bias the final error results. Therefore, direct comparison of the results of single-parabolic modeling with the results of dual-parabolic modeling may not be meaningful in the case of the MDCP errors. However, because the Hausdorff error measure is an indicator of the possibility of getting large errors on the average, a direct comparison of the Hausdorff errors for the results of the dual-parabolic modeling procedure to the Hausdorff errors for the results of the single-parabolic modeling procedures is acceptable.

Among the four different versions of the GHT, the two Gabor-magnitude-updated procedures have lower MDCP errors, on average, with and without the raphe-angle correction and the MDCP-based selection options being used, as compared to the unity-updated GHT procedures. The dual-parabolic modeling procedure for the ITA, using the Gabor-magnitude-updated

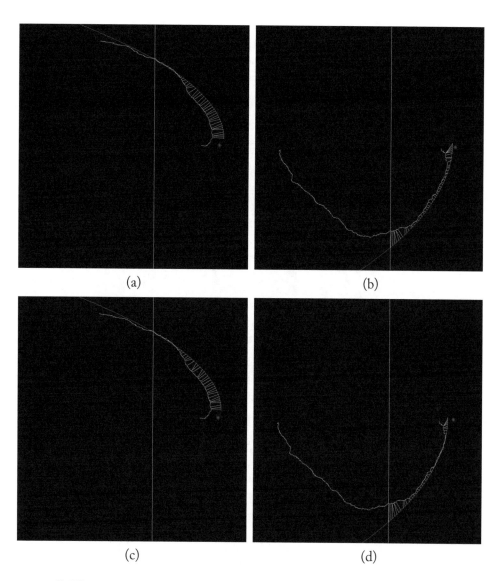

Figure 6.21: DCP errors for dual-parabolic modeling for image 6 of the DRIVE database, using the Gabor-magnitude-updated and vertex-restricted GHT with raphe-angle correction for (a) the STA and (b) the ITA models, resulting in MDCP errors of 15 and 11 pixels, respectively. (c) The STA and (d) the ITA models obtained using the Gabor-magnitude-updated and vertex-restricted GHT with MDCP-based selection and raphe-angle correction combined, resulting in MDCP errors of 14 and 9 pixels, respectively.

Table 6.2: The average MDCP errors (in pixels, where each pixel is 20 μm) of the parabolic fits for the four versions of the GHT, via the single- and dual-parabolic modeling procedures, as compared to the hand-drawn traces of the MTA for all of the 40 DRIVE images. The MDCP values for the procedures with raphe-angle correction, MDCP-based selection, and both combined are also provided. Note that VR stands for vertex restriction and GM stands for Gabor-magnitude

GHT Version	MDCP, Mean ± STD	With raphe-angle correction	With MDCP-based selection	With raphe-angle correction and MDCP-based selection
Unity-updated	18.35 ± 11.40	16.62 ± 9.42	16.26 ± 9.93	14.20 ± 7.07
Unity-updated with VR	16.27 ± 8.84	15.09 ± 7.85	15.15 ± 8.13	13.45 ± 7.54
GM-updated	14.08 ± 9.93	13.93 ± 9.20	12.68 ± 8.80	12.79 ± 8.63
GM-updated with VR	16.06 ± 9.05	12.64 ± 6.39	14.59 ± 8.00	12.10 ± 6.16
GM-updated with VR, ITA Model	12.07 ± 8.88	12.33 ± 11.02	10.90 ± 8.71	10.64 ± 8.76
GM-updated with VR, STA Model	15.01 ± 16.32	14.09 ± 15.28	14.52 ± 16.71	13.93 ± 16.06

Table 6.3: The average Hausdorff distance (in pixels, where each pixel is 20 μm) of the parabolic fits for the four versions of the GHT, via the single- and dual-parabolic modeling procedures, as compared to the hand-drawn traces of the MTA for all of the 40 DRIVE images. The Hausdorff distance values for the procedures with raphe-angle correction, MDCP-based selection, and both combined are also provided. Note that VR stands for vertex restriction and GM stands for Gabor-magnitude

GHT Version	Hausdorff, Mean ± STD	With raphe-angle correction	With MDCP-based selection	With raphe-angle correction and MDCP-based selection
Unity-updated	53.14 ± 34.42	49.17 ± 31.10	50.17 ± 34.27	42.50 ± 24.82
Unity-updated with VR	45.49 ± 22.33	40.94 ± 18.61	43.12 ± 20.78	36.75 ± 17.70
GM-updated	45.82 ± 34.52	46.05 ± 32.50	42.32 ± 29.29	43.55 ± 33.62
GM-updated with VR	46.72 ± 26.49	36.28 ± 15.63	42.51 ± 21.60	34.90 ± 16.60
GM-updated with VR, ITA Model	34.19 ± 15.70	35.25 ± 20.46	30.18 ± 15.73	29.80 ± 16.46
GM-updated with VR, STA Model	39.66 ± 31.51	38.06 ± 30.25	37.38 ± 32.27	36.36 ± 31.29

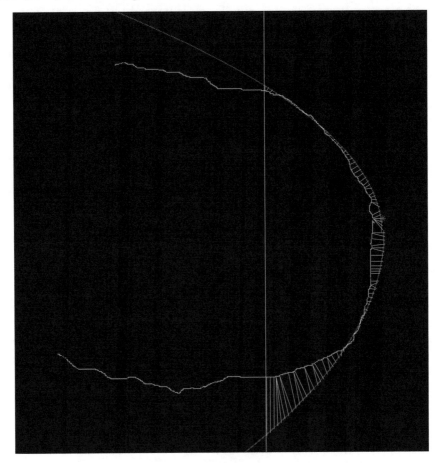

Figure 6.22: Illustration of the DCP errors for the result of dual-parabolic modeling for image 19 of the DRIVE database, using the Gabor-magnitude-updated GHT with vertex restriction. The MDCP error for the STA model is 5 pixels (0.11 mm). The MDCP error for the ITA model is 20 pixels (0.41 mm)

GHT with vertex restriction, has led to the lowest MDCP error, on average, with neither the raphe-angle correction nor the MDCP-based selection options being used. The results of the dual-parabolic modeling procedure for the STA have a higher MDCP error and larger STD, on average, as compared to the results of modeling the ITA, which can be attributed to the high MDCP errors produced by several oddly shaped STAs (outliers) in the DRIVE database. The added procedures of raphe-angle correction and MDCP-based selection have less impact when used separately; combining the two steps seems to have a bigger influence on the results of the unity-updated GHT, as compared to the Gabor-magnitude-updated GHT. The methods for raphe-angle correction and MDCP-based selection have a minor effect when applied to the dual-parabolic modeling procedure, as compared to the single-parabolic modeling procedures. The Hausdorff error measures (see Table 6.3) indicate that the dual-parabolic modeling procedure produces lower errors, on average, for both the ITA and the STA models, as compared to all four of the single-parabolic modeling procedures.

In order to test the statistical significance of the differences in the MDCP errors in the results provided by the various modeling options, the p-values for a few selected pairs of MDCP values were computed. The MDCP errors related to the Gabor-magnitude-updated GHT, with vertex restriction including raphe-angle correction and MDCP-based selection were found to be lower than those of the unity-updated GHT, with high statistical significance ($p < 0.01$). The differences between the MDCP errors of a few other pairs of options, such as the unity-updated GHT as compared with the Gabor-magnitude-updated GHT (without any other option), the Gabor-magnitude-updated GHT with and without vertex restriction, and Gabor-magnitude-updated GHT with and without all of the options listed in Table 6.2, were found to have no statistical significance.

6.9   PROGRAM RUN TIMES

Using a Lenovo Thinkpad T510, equipped with an Intel Core i7 (Hyper-threaded-dual-core) 2.67-GHz processor, 4 MB of level 2 cache, 8 GB of DDR3 RAM, running 64-bit Windows 7 Professional, and using 64-bit Matlab software, the run times for different parts of the procedures, used in the present work to model the MTA, for a single color image from the DRIVE database, are as follows:

- The Gabor-preprocessing step: 8.8 seconds

- The single-scale Gabor filtering step with $\{\tau, l, K\} = \{16, 2, 180\}$: 13.4 seconds

- The GHT-preprocessing step: 0.2 second

- The GHT step:

  1. Without any restriction: 18.4 seconds

  2. With the horizontal cropping option: 4.8 seconds

  3. With the vertex-restriction option: 14.6 seconds

  4. With the horizontal cropping and the vertex-restriction options: 4.5 seconds

- The GHT-postprocessing step with MDCP-based selection:

  1. Without any restriction: 3.8 seconds

  2. With the horizontal cropping option: 1.4 seconds

3. With the vertex-restriction option: 3.2 seconds

4. With the horizontal cropping and the vertex-restriction options: 1.2 seconds

The GHT-postprocessing steps, which select the global maximum in the Hough space, are generally about 0.5 second faster than the postprocessing steps that use the MDCP-based-selection option, as mentioned above. The horizontal limiting of the images has reduced the run time of the GHT algorithm by about 74%, as compared to the simplest form of the GHT procedure. Combining the vertex-restriction and the horizontal-cropping options reduces the run time by about 76%, as compared to the GHT process without any restrictions. On average, it takes about 30 seconds to perform all of the steps required for single-parabolic modeling of a color image from the DRIVE database. The dual-parabolic modeling procedure takes about the same time as the single-parabolic modeling procedure to run. The GHT run times could be lowered by a further reduction of the size of the Hough space. The size of the Hough space could be further reduced in the case of the vertex-restricted GHT, as there is only a square box with side $s = 40$ pixels that needs to be updated in the Hough space. The current size of the Hough space for the vertex-restricted GHT, $(584, 181, 86)$, could be reduced to $(40, 40, 86)$ with additional modifications to the indexing and addressing procedures. Sparse matrices could also be used to reduce the storage requirements of the Hough space. In addition, it is easily possible to translate all of the procedures mentioned above to C++ and consequently, to hardware programming languages that are similar to C++ in syntax, such as DSP++ [188], for more efficient and faster implementation of the proposed methods. The results in this chapter indicate that the methods, as proposed in the present work, are suitable for implementation in a clinical setting, as discussed in Chapter 7.

6.10 DISCUSSION

The addition of the anatomical restrictions on the vessel map and the search area in the Hough space, as well as the use of Gabor-magnitude-weighted increments, have significantly improved the results of parabolic modeling. By limiting the search area for the vertex to be close to the automatically detected center of the ONH, the GHT procedure is forced to fit the parabolic model closer to the posterior part of the MTA. By restricting the vessel map up to the macula, the parabolic profile of the appropriate portion of the MTA is emphasized. Updating the Hough space with the Gabor magnitude response reduces the influence of smaller vessels on the result. This is confirmed by the lower average MDCP values for the results of the Gabor-magnitude-updated GHT, as compared to those of the unity-updated GHT (see Tables 6.2 and 6.3).

The high correlation between the values of the parameters a obtained automatically and from the hand-drawn MTAs indicates that the automatically obtained VSTs are accurate.

The MDCP-based selection procedure, applied to the raphe-angle-corrected images, helps to improve the accuracy of the models. The addition of the procedures for raphe-angle correction and MDCP-based selection to the single-parabolic modeling GHT procedure appears to have the same impact as the dual-parabolic modeling procedure, without the use of raphe-angle correction and MDCP-based selection. The addition of raphe-angle correction and MDCP-based selection has a minor impact on the results of the dual-parabolic modeling procedure. The dual-parabolic modeling procedure is more reliable, as it is less likely to result in a large error, as shown by the lower Hausdorff errors, on average, as compared to the single-parabolic modeling procedures.

Parabolic modeling, with conditions as described in the present chapter, characterizes the architecture of the MTA up to the macular region, whereas the procedure of Wilson et al. [32] quantifies an angle based on the location of the ONH and only two specific points on the ITA and STA that reflect the location of the fovea (see Figure 1.7). This observation implies that the two measures are fundamentally different from each other; this point is confirmed by low correlation values obtained between the automatically measured values of the TAA, the IAA, and the SAA, using the procedure described by Wilson et al. [32], and the automatically obtained openness parameters, a_{MTA}, a_{ITA}, and a_{STA}, respectively, as shown in Table 6.4. As illustrated by Ells and MacKeen [49], analysis of the changes to the entire architecture of the MTA may be more desirable than analysis of the variations in the TAA in the presence of progressive ROP.

Using a thinning algorithm, as opposed to a curvature-skeleton algorithm (see Section 3.1.4), to obtain the skeleton of the MTA could be more beneficial when using the GHT procedure, as a thinning algorithm ignores small branching points and emphasizes the curvilinear structure of the MTA. Additional procedures to select parts of the MTA without the presence of other vessels can lead to improved performance of the modeling methods.

6.11 REMARKS

The modeling methods were tested on the DRIVE images in the present work; however, the DRIVE database neither has information regarding the TAA, nor does it contain information regarding possible diseases that affect the angle of insertion of the MTA. Regardless, the DRIVE database is one of a few public databases that is commonly used in the design and evaluation of algorithms for the analysis of retinal fundus images. The DRIVE database has served the main purpose of the present work, which is to develop image processing methods for the detection and modeling of the MTA, as well as their assessment using a well-established database of retinal images.

Retinal fundus images of preterm infants typically lack a clear depiction of the fovea; as shown by Chiang et al. [189], there is significant variability in the identification of the fovea in wide-angle retinal images of preterm infants at risk of ROP, even between ROP experts. This fact may preclude the application of the method of Wilson et al. [32] for the measurement of the TAA, which is dependent on the location of the fovea. The methods described in the present chapter require only the approximate location of the center of the ONH, and thereby provide the advantage of requiring a single landmark.

Potential clinical applications of the modeling methods are presented in Chapter 7.

Table 6.4: The correlation coefficient values obtained between the automatically obtained values of the openness parameters a_{MTA}, a_{STA}, and a_{ITA}, and the automatically obtained values of the TAA, SAA, and IAA, respectively. Values for the TAA, SAA, and IAA were computed by using manual markings of the ONH and fovea, as well as the hand-drawn traces of the MTA, based on the procedure described by Wilson et al. [40]. The higher correlation values for the cases where the raphe-angle-correction step has been used can be attributed to the fact that the procedure by Wilson et al. [40] requires the raphe angle to be corrected before measuring the SAA and IAA values. In general, the values of the openness parameter a, are not well correlated to values of the angle of insertion of the MTA, measured as described by Wilson et al. [40]. Note that VR stands for vertex restriction and GM stands for Gabor-magnitude

GHT Version	Correlation coefficient without any options	With raphe-angle correction	With MDCP-based selection	With raphe-angle correction and MDCP-based selection
Unity-updated	0.2476	0.1583	0.1319	0.2902
Unity-updated with VR	0.4303	0.4713	0.4901	0.4345
GM-updated	0.3523	0.4735	0.4744	0.4556
GM-updated with VR	0.1563	0.6301	0.3676	0.6558
GM-updated with VR, ITA Model	0.4421	0.6433	0.5730	0.6326
GM-updated with VR, STA Model	0.3108	0.4738	0.2704	0.4236

CHAPTER 7

Potential Clinical Applications

In order to facilitate the clinical application of methods for the detection of blood vessels (see Chapter 5) and modeling of the MTA (see Chapter 6), a graphical user interface (GUI) was developed [190]. Figure 7.1 shows a screen-shot of the GUI. The GUI is deployed as a stand-alone application package, which can be installed on Windows operating systems and does not require a MATLAB license to run.

As mentioned in Section 1.2.1, one of the pathologies that can be diagnosed by analyzing the openness of the MTA is PDR. Diagnosis of PDR via the GUI was performed in a study [47] using 11 retinal images showing signs of PDR and 11 images of normal cases obtained from the STARE database [169] (see Section 4.1.2). The results obtained using the openness parameter of the dual-parbolic model indicated an area under the ROC curve of $A_z = 0.87$, in terms of the diagnosis of PDR versus normal cases [47]. Figure 7.2 shows examples of normal and PDR cases, as well as the results of dual-parabolic modeling for each case. It is clear, by comparison, that the MTA becomes narrower in the presence of PDR.

The presence of plus disease, as mentioned in Section 1.2.2, is also known to reduce the openness of the MTA. The proposed GUI was used along with 110 images from the TROPIC database [171] (see Section 4.1.3) to perform the diagnosis of plus cases versus normal cases by quantification of the openness of the MTA [37]. Using the results of dual-parabolic modeling, an average area under the ROC curve of $A_z = 0.80$ was achieved in the diagnosis of plus disease with a set of 19 normal images and 19 images of patients with plus disease. The results indicate the suitability of the methods for the detection and modeling of the MTA to perform diagnosis of plus disease. Figure 7.3 shows two images from the TROPIC database along with their corresponding dual-parabolic models. Figure 7.3 (b) is a normal case, whereas Figure 7.3 (d) is a case with plus disease. It is evident, by visual and quantitative comparison, that the openness of the MTA reduces in the presence of plus disease.

The methods developed in the present work for the detection of blood vessels have achieved the highest detection rates ($A_z = 0.96$) among the methods reported in the literature that have been tested with the DRIVE database, as explained in Section 5.6. Real Gabor filters are suitable for the detection of oriented and branching patterns, in general, and are not limited to use in a biomedical application, as presented in this book. Similarly, the methods proposed for the modeling of the MTA (the GHT) are suitable for the modeling of similar geometrical patterns in images regardless of the application.

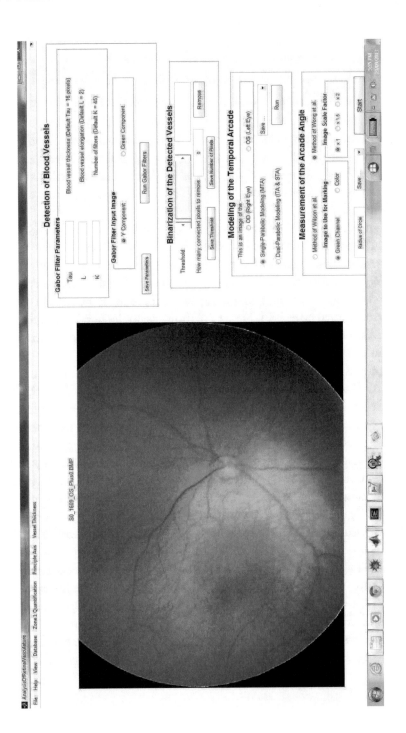

Figure 7.1: A GUI being developed to incorporate the procedures for the detection and parabolic modeling of retinal blood vessels.

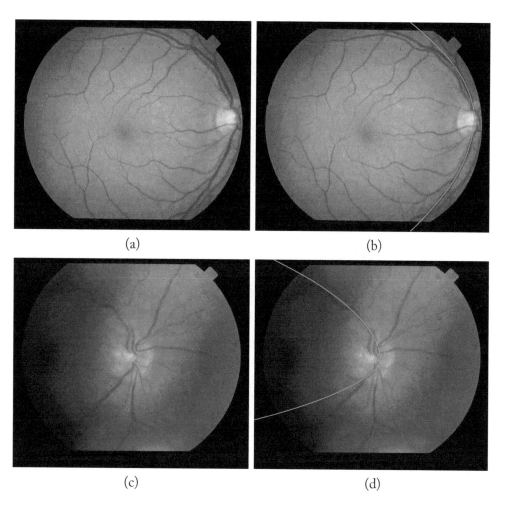

Figure 7.2: (a) Image 0082 of the STARE database, which is a normal case. (b) Results of dual-parabolic modeling of (a), with $a_{STA} = -108$ and $a_{ITA} = -156$. (c) Image 0348 of the STARE database, which is a PDR case. (d) Results of dual-parabolic modeling of (c), with $a_{STA} = -72$ and $a_{ITA} = -27$. The inferior model is fitting the ITA close to the ONH. It is clear, both visually and quantitatively, that the openness of the MTA in the presence of PDR is significantly reduced, as compared to the normal case.

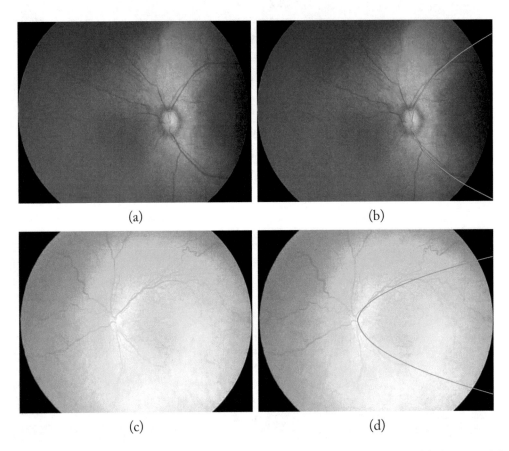

(a) (b)

(c) (d)

Figure 7.3: (a) Image 2001 of the TROPIC database, which is a normal case. (b) Results of dual-parabolic modeling of (a), with $a_{STA} = 66$ and $a_{ITA} = 48$. (c) Image 2903 of the TROPIC database, which is of a patient with stage 3 ROP as well as plus disease. (d) Results of dual-parabolic modeling of (c), with $a_{STA} = 20$ and $a_{ITA} = 28$. It is evident, both by visual and quantitative comparison, that the openness of the MTA in the presence of ROP and plus disease is reduced, as compared to the normal case.

References

[1] Michaelson IC and Benezra D. *Textbook of the Fundus of the Eye*. Churchill Livingstone, Edinburgh, UK, 3rd edition, 1980. 1, 18, 19

[2] Hansen AB, Sander B, Larsen M, Kleener J, Borch-Johnsen K, and Lund-Andersen H. Screening for diabetic retinopathy using a digital non-mydriatic camera compared with standard 35-mm stereo colour transparencies. *Acta Ophthalmologica Scandinavica*, 82(6):656–665, December 2004. DOI: 10.1111/j.1600-0420.2004.00347.x. 1

[3] Klein B, Klein R, Hall E, Lee K, and Jensen K. The compatibility of estimates of retroil-luminated lens opacities as judged from film-based and digital imaging. *American Journal of Ophthalmology*, 138:668–670, 2004. DOI: 10.1016/j.ajo.2004.04.068.

[4] van Leeuwen R, Chakravarthy U, Vingerling JR, Brussee C, Hooghart A, Mudler P, and de Jong P. Grading of age-related maculopathy for epidemiological studies: Is digital imaging as good as 35-mm film? *Ophthalmology*, 110(8):1540–1544, 2003. DOI: 10.1016/S0161-6420(03)00501-3. 1

[5] Patton N, Aslam TM, MacGillivray T, Deary IJ, Dhillon B, Eikelboom RH, Yogesan K, and Constable IJ. Retinal image analysis: Concepts, applications and potential. *Progress in Retinal and Eye Research*, 25(1):99–127, 2006. DOI: 10.1016/j.preteyeres.2005.07.001. 1, 5, 7, 9, 11, 13, 15, 67

[6] Chapman N, Witt N, Gao X, Bharath AA, Stanton AV, Thom SA, and Hughes AD. Computer algorithms for the automated measurement of retinal arteriolar diameters. *British Journal of Ophthalmology*, 85:74–79, 2001. DOI: 10.1136/bjo.85.1.74.

[7] Swanson C, Cocker KD, Parker KH, Moseley MJ, and Fielder AR. Semiautomated computer analysis of vessel growth in preterm infants without and with ROP. *British Journal of Ophthalmology*, 87(12):1474–1477, 2003. DOI: 10.1136/bjo.87.12.1474. 5, 8, 9, 13, 67

[8] Eze CU, Gupta R, and Newman DL. A comparison of quantitative measures of arterial tortuosity using sine wave simulations and 3D wire models. *Physics in Medicine and Biology*, 45:2593–2599, 2000. DOI: 10.1088/0031-9155/45/9/312.

[9] Narasimha-Iyer H, Can A, Roysam B, Stewart CV, Tanenbaum HL, Majerovics A, and Singh H. Robust detection and classification of longitudinal changes in color retinal fundus images for monitoring diabetic retinopathy. *IEEE Transactions on Biomedical Engineering*, 53(6):1084–1098, 2006. DOI: 10.1109/TBME.2005.863971. 13, 15

[10] Walter T, Klein JC, Massin P, and Erginay A. A contribution of image processing to the diagnosis of diabetic retinopathy—detection of exudates in color fundus images of the human retina. *IEEE Transactions on Medical Imaging*, 21(10):1236–1243, 2002. DOI: 10.1109/TMI.2002.806290. 1

[11] Goldbaum M, Hart WE, Côté BL, and Raphaelian PV. Automated measures of retinal blood vessel tortuosity. *Investigative Ophthalmology and Visual Science*, 35:2089, 1994. 1

[12] Hart WE, Goldbaum M, Côté BL, Kube P, and Nelson MR. Automated measurement of retinal vascular tortuosity. In *Proceedings of the American Medical Informatics Association Annual Fall Conference*, pages 459–463, 1997.

[13] Hart WE, Goldbaum M, Côté B, Kube P, and Nelson MR. Measurement and classification of retinal vascular tortuosity. *International Journal of Medical Informatics*, 53(2-3):239–252, 1999. DOI: 10.1016/S1386-5056(98)00163-4. 1, 8, 13

[14] Staal JJ, Abramoff MD, Niemeijer M, Viergever MA, and van Ginneken B. Ridge based vessel segmentation in color images of the retina. *IEEE Transactions on Medical Imaging*, 23(4):501–509, 2004. DOI: 10.1109/TMI.2004.825627. 1, 57

[15] Bron AJ, Tripathi RC, and Tripathi BJ. *Wolff's Anatomy of the Eye and Orbit*. Arnold, London, UK, eighth edition, 2001. 2, 4

[16] Levine MD. *Vision in Man and Machine*. McGraw-Hill, New York, NY, 1985. 3

[17] Marr D. *Vision: A Computational Investigation into the Human Representation and Processing of Visual Information*. WH Freeman, San Francisco, CA, 1982. 3

[18] Snell SR and Lemp AM. *Clinical Anatomy of the Eye*. Blackwell Science, New York, 2nd edition, 1998. 3

[19] Larsen HW. *The Ocular Fundus: A Color Atlas*. Munksgaard, Copenhagen, Denmark, 1976. 3, 5, 19, 97, 99

[20] Lalonde M, Beaulieu M, and Gagnon L. Fast and robust optic disc detection using pyramidal decomposition and Hausdorff-based template matching. *IEEE Transactions on Medical Imaging*, 20(11):1193–1200, 2001. DOI: 10.1109/42.963823. 3, 19, 99

[21] Evans J, Rooney C, Ashgood S, Dattan N, and Wormald R. Blindness and partial sight in England and Wales April 1900–March 1991. *Health Trends*, 28:5–12, 1996. 5

[22] Fong DS, Aiello L, Gardner TW, King GL, Blankenship G, Cavallerano JD, Ferris FL, and Klein R. Retinopathy in diabetes. *Diabetes Care*, 27:84–87, 2004. DOI: 10.2337/diacare.27.10.2540.

[23] Noble J and Chaudhary V. Diabetic retinopathy. *Canadian Medical Association Journal*, 182:1646–1646, 2010. DOI: 10.1503/cmaj.090536. 5

[24] Jelinek HF and Cree MJ. Introduction. In Jelinek HF and Cree MJ, editors, *Automated Image Detection of Retinal Pathology*, pages 1–26. CRC Press, Boca Raton, FL, 2010. 5, 7, 9, 13

[25] Kohner E and Sleightholm M. Does microaneurysm count reflect the severity of the early diabetic retinopathy. *Opththalmology*, 93(5):586–589, 1986. DOI: 10.1016/S0161-6420(86)33692-3. 5

[26] Klein R, Meuer SM, and Moss SE. Retinal microaneurysm counts and 10-year progression of diabetic retinopathy. *Archives of Ophthalmology*, 113(11):1386–1391, 1995. DOI: 10.1001/archopht.1995.01100110046024. 5

[27] Meier P and Wiedemann P. Vitrectomy for traction macular detachment in diabetic retinopathy. *Graefe's Archive for Clinical and Experimental Ophthalmology*, 235:569–574, 1997. DOI: 10.1007/BF00947086. 7

[28] Danis RP and Davis MD. Proliferative diabetic retinopathy. In Duh EJ, editor, *Diabetic Retinopathy*, Contemporary Diabetes, pages 29–65. Humana Press, Totowa, NJ, 2008. DOI: 10.1007/978-1-59745-563-3. 7

[29] Meyerle CB, Chew EY, and FL Ferris III. Nonproliferative diabetic retinopathy. In Duh EJ, editor, *Diabetic Retinopathy*, Contemporary Diabetes, pages 3–27. Humana Press, Totowa, NJ, 2008. DOI: 10.1007/978-1-59745-563-3. 7

[30] International Committee for the Classification of Retinopathy of Prematurity. The international classification of retinopathy of prematurity revisited. *Archives of Ophthalmology*, 123:991–999, 2005. DOI: 10.1001/archopht.123.7.991. 7, 8, 9, 10

[31] Cryotherapy for Retinopathy of Prematurity Cooperative Group. Multicenter trial of cryotherapy for retinopathy of prematurity: Ophthalmological outcomes at 10 years. *Archives of Ophthalmology*, 119:1110–1118, 2001. DOI: 10.1001/archopht.119.8.1110. 7, 10

[32] Wilson CM, Cocker KD, Moseley MJ, Paterson C, Clay ST, Schulenburg WE, Mills MD, Ells AL, Parker KH, Quinn GE, Fielder AR, and Ng J. Computerized analysis of retinal vessel width and tortuosity in premature infants. *Investigative Ophthalmology and Visual Science*, 49(1):3577–3585, 2008. DOI: 10.1167/iovs.07-1353. 7, 9, 125

[33] Watzke RC, Robertson JE, Palmer EA, Wallace PR, Evans MS, and Delaney Soldevilla JE. Photographic grading in the retinopathy of prematurity cryotherapy trial. *Archives of Ophthalmology*, 108(7):950–955, 1990. DOI: 10.1001/archopht.1990.01070090052038. 7, 8

[34] Wallace DK, Jomier J, Aylward SR, and Landers, III MB. Computer-automated quantification of plus disease in retinopathy of prematurity. *Journal of American Association for Pediatric Ophthalmology and Strabismus*, 7:126–130, April 2003. DOI: 10.1067/mpa.2003.S1091853102000150. 7

[35] Gelman R, Martinez-Perez ME, Vanderveen DK, Moskowitz A, and Fulton AB. Diagnosis of plus disease in retinopathy of prematurity using retinal image multiscale analysis. *Investigative Ophthalmology & Visual Science*, 46(12):4734–4738, 2005. DOI: 10.1167/iovs.05-0646. 7, 9

[36] Wallace DK, Kylstra JA, and Chesnutt DA. Prognostic significance of vascular dilation and tortuosity insufficient for plus disease in retinopathy of prematurity. *American Association for Pediatric Ophthalmology and Strabismus*, 4(4):224–229, 2000. DOI: 10.1067/mpa.2000.105273. 7, 9

[37] Oloumi F, Rangayyan RM, and Ells AL. Quantitative analysis of the major temporal arcade in retinal fundus images of preterm infants for detection of plus disease. In *Proc. IASTED International Conference on Signal and Image Processing*, pages 464–469, Banff, Alberta, Canada, July 2013. DOI: 10.2316/P.2013.804-047. 7, 13, 127

[38] Freedman SF, Kylstra JA, Capowski JJ, Realini TD, Rich C, and Hunt D. Observer sensitivity to retinal vessel diameter and tortuosity in retinopathy of prematurity. *Journal of Pediatric Ophthalmology and Strabismus*, 33:248–254, 1996. 7

[39] Wallace DK, Quinn GE, Freedman SF, and Chiang MF. Agreement among pediatric ophthalmologists in diagnosing plus and pre-plus disease in retinopathy of prematurity. *Journal of American Association for Pediatric Ophthalmology and Strabismus*, 12(4):352–356, 2008. DOI: 10.1016/j.jaapos.2007.11.022. 7, 13

[40] Wilson C, Theodorou M, Cocker KD, and Fielder AR. The temporal retinal vessel angle and infants born preterm. *British Journal of Ophthalmology*, 90:702–704, 2006. DOI: 10.1136/bjo.2005.085019. 8, 9, 10, 65, 108, 126

[41] Fledelius HC and Goldschmidt E. Optic disc appearance and retinal temporal vessel arcade geometry in high myopia, as based on follow-up data over 38 years. *Acta Ophthalmologica*, 88(5):514–520, 2010. DOI: 10.1111/j.1755-3768.2009.01660.x. 8, 9, 10, 13, 20

[42] Reese AB, King MJ, and Owens WC. A classification of retrolental fibroplasia. *American Journal of Ophthalmology*, 36:1333–1335, 1953. 8, 10

[43] Heneghan C, Flynn J, O'Keefe M, and Cahill M. Characterization of changes in blood vessels width and tortuosity in retinopathy of prematurity using image analysis. *Medical Image Analysis*, 6(1):407–429, 2002. DOI: 10.1016/S1361-8415(02)00058-0. 8, 9, 13

[44] Shah DN, Karp KA, Ying G, Mills MD, and Quinn GE. Image analysis of posterior pole vessels identifies type 1 retinopathy of prematurity. *American Association for Pediatric Ophthalmology and Strabismus*, 13(5):507–508, 2009. DOI: 10.1016/j.jaapos.2009.06.007.

[45] Shah DN, Wilson CM, Ying G, Karp KA, Fielder AR, Ng J, Mills MD, and Quinn GE. Semiautomated digital image analysis of posterior pole vessels in retinopathy of prematurity. *American Association for Pediatric Ophthalmology and Strabismus*, 13(5):504–506, 2009. DOI: 10.1016/j.jaapos.2009.06.007. 8, 13

[46] Tokoro T. *Atlas of Posterior Fundus Changes in Pathological Myopia*. Springer-Verlag, Tokyo, Japan, 1998. DOI: 10.1007/978-4-431-67951-6. 9

[47] Oloumi F, Rangayyan RM, and Ells AL. Computer-aided diagnosis of proliferative diabetic retinopathy via modeling of the major temporal arcade in retinal fundus images. *Journal of Digital Imaging*, 26(6):1124–1130, December 2013. DOI: 10.1007/s10278-013-9592-9. DOI: 10.1007/s10278-013-9592-9. 9, 13, 127

[48] Wong K, Ng J, Ells AL, Fielder AR, and Wilson CM. The temporal and nasal retinal arteriolar and venular angles in preterm infants. *British Journal of Ophthalmology*, 95(12):1723–1727, 2011. DOI: 10.1136/bjophthalmol-2011-300416. 9, 10

[49] Ells AL and MacKeen LD. Retinopathy of prematurity—the movie. *Journal of American Association for Pediatric Ophthalmology and Strabismus*, 8(4):389, 2004. DOI: 10.1016/j.jaapos.2003.08.014. 10, 125

[50] Tuli D and Camras CB. Glaucoma. In Gendelman HE and Ikezu T, editors, *Neuroimmune Pharmacology*. Springer, New York, 2008. DOI: 10.1007/978-0-387-72573-4. 11

[51] Noble J and Chaudhary V. Age-related macular degeneration. *Canadian Medical Association Journal*, 182(16):1759–1759, 2010. DOI: 10.1503/cmaj.090378. 11

[52] Wong TY, Shankar A, and Klein R. Retinal arteriolar narrowing, hypertension and subsequent risk of diabetes mellitus. *Archives of Internal Medicine*, 165(9):1060–1065, 2005. DOI: 10.1001/archinte.165.9.1060. 11

[53] Witt N, Wong TY, Hughes AD, Chaturvedi N, Klein BE, Evans R, McNamara M, McG Thom SA, and Klein R. Abnormalities of retinal microvascular structure and risk of mortality from ischemic heart disease and stroke. *Hypertension*, 47(5):975–981, May 2006. DOI: 10.1161/01.HYP.0000216717.72048.6c. 11

[54] KIDROP: Karnataka Internet Assisted Diagnosis of Retinopathy of Prematurity. `http://www.narayananethralaya.org/kidrop.html`. 11

[55] Fijalkowski N, Zheng LL, Henderson MT, Wallenstein MB, Leng T, and Moshfeghi DM. Stanford University Network for Diagnosis of Retinopathy of Prematurity (SUNDROP): Four-years of screening with telemedicine. *Current Eye Research*, 38(2):283–291, 2013. DOI: 10.3109/02713683.2012.754902. 11

[56] Abràmoff MD and Niemeijer M. Detecting retinal pathology automatically with special emphasis on diabetic retinopathy. In Jelinek HF and Cree MJ, editors, *Automated Image Detection of Retinal Pathology*, pages 67–78. CRC Press, Boca Raton, FL, 2010. 13

[57] Bäcklund LB. Finding a role for computer-aided early diagnosis of diabetic retinopathy. In Jelinek HF and Cree MJ, editors, *Automated Image Detection of Retinal Pathology*, page 79. CRC Press, Boca Raton, FL, 2010.

[58] Wang H, Hsu W, Guan K, and Lee M. An effective approach to detect lesions in color retinal images. In *Proceedings of the IEEE Conference on Computer Vision and Pattern Recognition*, pages 181–186, Hilton Head Island, SC, 2000. IEEE. DOI: 10.1109/CVPR.2000.854775.

[59] Sopharak A and Uyyanonvara B. Automatic exudates detection from diabetic retinopathy retinal image using fuzzy c-means and morphological methods. In *Proceedings of the Third IASTED International Conference: Advances in Computer Science and Technology*, pages 359–364, Anaheim, CA, 2007.

[60] Quellec G, Lamard M, Josselin PM, Cazuguel G, Cochener B, and Roux C. Optimal wavelet transform for the detection of microaneurysms in retina photographs. *IEEE Transactions on Medical Imaging*, 27(9):1230–1241, September 2008. DOI: 10.1109/TMI.2008.920619.

[61] Xiaohui Z and Chutatape A. Detection and classification of bright lesions in color fundus images. In *Image Processing, 2004. ICIP '04. 2004 International Conference on*, volume 1, pages 139–142, October 2004. DOI: 10.1109/ICIP.2004.1418709.

[62] Niemeijer M, Abràmoff MD, and van Ginneken B. Information fusion for diabetic retinopathy CAD in digital color fundus photographs. *IEEE Transactions on Medical Imaging*, 28(5):775–785, 2009. DOI: 10.1109/TMI.2008.2012029.

[63] Kande GB, Savithri TS, and Subbaiah PV. Automatic detection of microaneurysms and hemorrhages in digital fundus images. *Journal of Digital Imaging*, 23(4):430–437, August 2010. DOI: 10.1007/s10278-009-9246-0.

[64] Gregson PH, Shen Z, Scott RC, and Kozousek V. Automated grading of venous beading. *Computers and Biomedical Research*, 28:291–304, August 1995. DOI: 10.1006/cbmr.1995.1020.

[65] Smith RT, Chan JK, Nagasaki T, Ahmed UF, Barbazetto I, Sparrow J, Figueroa M, and Merriam J. Automated detection of macular drusen using geometric background leveling and threshold selection. *Archives of Ophthalmology*, 123:200–206, 2005. DOI: 10.1001/archopht.123.2.200.

[66] Köse C, Şvik U, and Gençalioğlu O. Automatic segmentation of age-related macular degeneration in retinal fundus images. *Computers in Biology and Medicine*, 38:611–619, 2008. DOI: 10.1016/j.compbiomed.2008.02.008.

[67] Friberg TR, Huang L, Palaiou M, and Bremer R. Computerized detection and measurement of drusen in age-related macular degeneration. *Ophthalmic Surgery, Lasers and Imaging*, 38:126–134, 2007. 13

[68] Zhang Z, Yin FS, Liu J, Wong WK, Tan NM, Lee BH, Cheng J, and Wong TY. ORIGA^{light}: An online retinal fundus image database for glaucoma analysis and research. In *Engineering in Medicine and Biology Society, 32nd Annual International Conference of the IEEE*, pages 3065–3068, Buenos Aires, Argentina, 2010. DOI: 10.1109/IEMBS.2010.5626137. 13

[69] Nayak J. Automated diagnosis of glaucoma using digital fundus images. *Journal of Medical Systems*, 33:337–346, 2009. DOI: 10.1007/s10916-008-9195-z.

[70] Muramatsu C, Hayashi Y, Sawada A, Hatanaka Y, Hara T, Yamamoto T, and Fujita H. Detection of retinal nerve fiber layer defects on retinal fundus images for early diagnosis of glaucoma. *Journal of Biomedical Optics*, 15:16–21, 2010. DOI: 10.1117/1.3322388. 13

[71] Chaudhuri S, Chatterjee S, Katz N, Nelson M, and Goldbaum M. Detection of blood vessels in retinal images using two-dimensional matched filters. *IEEE Transactions on Medical Imaging*, 8(3):263–269, 1989. DOI: 10.1109/42.34715. 15, 23, 35

[72] Fleming AD, Goatman KA, Philip S, Olson JA, and Sharp PF. Automatic detection of retinal anatomy to assist diabetic retinopathy screening. *Physics in Medicine and Biology*, 52:331–345, 2007. DOI: 10.1088/0031-9155/52/24/012. 15, 19, 20, 93, 94

[73] Hoover A and Goldbaum M. Locating the optic nerve in a retinal image using the fuzzy convergence of the blood vessels. *IEEE Transactions on Medical Imaging*, 22(8):951–958, August 2003. DOI: 10.1109/TMI.2003.815900. 15, 19

[74] Foracchia M, Grisan E, and Ruggeri A. Detection of optic disc in retinal images by means of a geometrical model of vessel structure. *IEEE Transactions on Medical Imaging*, 23(10):1189–1195, 2004. DOI: 10.1109/TMI.2004.829331. 15, 18, 19

[75] Lowell J, Hunter A, Steel D, Basu A, Ryder R, and Kennedy RL. Measurement of retinal vessel widths from fundus images based on 2-D modeling. *IEEE Transactions on Medical Imaging*, 23(10):1196–1204, 2004. DOI: 10.1109/TMI.2004.830524. 18

[76] Hoover A, Kouznetsova V, and Goldbaum M. Locating blood vessels in retinal images by piecewise threshold probing of a matched filter response. *IEEE Transactions on Medical Imaging*, 19(3):203–210, 2000. DOI: 10.1109/42.845178. 16, 23, 35

[77] Staal J, Abràmoff MD, Niemeijer M, Viergever MA, and van Ginneken B. Ridge-based vessel segmentation in color images of the retina. *IEEE Transactions on Medical Imaging*, 23(4):501–509, 2004. DOI: 10.1109/TMI.2004.825627. 16, 18, 23, 35

[78] Li H and Chutatape O. Automated feature extraction in color retinal images by a model based approach. *IEEE Transactions on Biomedical Engineering*, 51(2):246–254, 2004. DOI: 10.1109/TBME.2003.820400. 15, 19

[79] Niemeijer M, Staal J, van Ginneken B, Loog M, and Abràmoff MD. Comparative study of retinal vessel segmentation methods on a new publicly available database. In *Proceedings of the SPIE International Symposium on Medical Imaging*, pages 648–656. SPIE, 2004. DOI: 10.1117/12.535349.

[80] Lalonde M, Laliberté F, and Gagnon L. RetsoftPlus: A tool for retinal image analysis. In *Proceedings of the 17th IEEE Symposium on Computer-based Medical Systems*, pages 542–547. IEEE Computer Society, 2004. DOI: 10.1109/CBMS.2004.1311771. 15

[81] Tobin KW, Chaum E, Govindasamy VP, and Karnowski TP. Detection of anatomic structures in human retinal imagery. *IEEE Transactions on Medical Imaging*, 26(12):1729–1739, December 2007. DOI: 10.1109/TMI.2007.902801. 15, 19, 21

[82] Kochner B, Schuhmann D, Michaelis M, Mann G, and Englmeier KH. Course tracking and contour extraction of retinal vessels from color fundus photographs: most efficient use of steerable filters for model based image analysis. In *SPIE Medical Imaging*, volume 3338, pages 755–761, San Diego, CA, February, 1998. DOI: 10.1117/12.310955. 19, 93, 94

[83] Ying H and Liu JC. Automated localization of macula-fovea area on retina images using blood vessel network topology. In *Acoustics Speech and Signal Processing, IEEE International Conference on*, pages 650–653, March 2010. DOI: 10.1109/ICASSP.2010.5495144. 15, 19, 20

[84] Jiang X and Mojon D. Adaptive local thresholding by verification-based multi-threshold probing with application to vessel detection in retinal images. *IEEE Transactions on Pattern Analysis and Machine Intelligence*, 25(1):131–137, 2003. DOI: 10.1109/TPAMI.2003.1159954. 15

[85] Soares JVB, Leandro JJG, Cesar Jr. RM, Jelinek HF, and Cree MJ. Retinal vessel segmentation using the 2-D Gabor wavelet and supervised classification. *IEEE Transactions on Medical Imaging*, 25(9):1214–1222, 2006. DOI: 10.1109/TMI.2006.879967. 16, 18, 23, 36, 72, 89

[86] Rangayyan RM, Ayres FJ, Oloumi Faraz, Oloumi Foad, and Eshghzadeh-Zanjani P. Detection of blood vessels in the retina with multiscale Gabor filters. *Journal of Electronic Imaging*, 17:023018:1–7, April-June 2008. DOI: 10.1117/1.2907209. 16, 36, 67, 72, 89

[87] Al-Diri B, Hunter A, and Steel D. An active contour model for segmenting and measuring retinal vessels. *IEEE Transactions on Medical Imaging*, 28(9):1488–1497, September 2009. DOI: 10.1109/TMI.2009.2017941. 16

[88] Ushizima DM, Medeiros FNS, Cuadros J, and Martins CIO. Vessel network detection using contour evolution and color components. In *Engineering in Medicine and Biology Society, 32nd Annual International Conference of the IEEE*, pages 3129–3132, Buenos Aires, Argentina, September 2010. DOI: 10.1109/IEMBS.2010.5626090. 17

[89] Marin D, Aquino A, Gegundez-Arias ME, and Bravo JM. A new supervised method for blood vessel segmentation in retinal images by using gray-level and moment invariants-based features. *IEEE Transactions on Medical Imaging*, 30(1):146–158, January 2011. DOI: 10.1109/TMI.2010.2064333. 17

[90] Frangi AF, Niessen WJ, Vincken KL, and Viergever MA. Multiscale vessel enhancement filtering. In *Medical Image Computing and Computer-Assisted Intervention - MICCAI'98*, volume 1496 of *Lecture Notes in Computer Science*, pages 130–137. Springer, Berlin, Germany, 1998. DOI: 10.1007/BFb0056195. 17, 18

[91] Dhara AK, Rangayyan RM, Oloumi F, and Mukhopadhyay S. Methods for the detection of blood vessels in retinal fundus images and reduction of false-positive pixels around the optic nerve head. In *Proceedings of the 4th IEEE International Conference on E-Health and Bioengineering - EHB 2013*, pages 1–6, Iaşi, Romania, November 2013. DOI: 10.1109/EHB.2013.6707365. 17, 89

[92] Lupaşcu CA, Tegolo D, and Trucco E. FABC: Retinal vessel segmentation using AdaBoost. *IEEE Transactions on Information Technology in Biomedicine*, 14(5):1267–1274, September 2010. DOI: 10.1109/TITB.2010.2052282. 17

[93] Lindeberg T. Edge detection and ridge detection with automatic scale selection. *International Journal of Computer Vision*, 30(2):117–154, 1998. DOI: 10.1023/A:1008097225773. 18

[94] Sofka M and Stewart CV. Retinal vessel centerline extraction using multiscale matched filters, confidence and edge measures. *IEEE Transactions on Medical Imaging*, 25(12):1531–1546, December 2006. DOI: 10.1109/TMI.2006.884190. 18

[95] Freund Y and Schapire R. A short introduction to boosting. *Japanese Society for Artificial Intelligence, Journal of*, 14(5):771–780, September 1999. 18

[96] Lupaşcu CA, Tegolo D, and Trucco E. A comparative study on feature selection for retinal vessel segmentation using FABC. In Jiang X and Petkov N, editors, *Computer Analysis of Images and Patterns*, volume 5702, pages 655–662. Springer, Berlin, Germany, 2009. 18

[97] Lam BSY, Gao Y, and Liew AW-C. General retinal vessel segmentation using regulization-based multiconcavity modeling. *IEEE Transactions on Medical Imaging*, 29(7):1369–1381, July 2010. DOI: 10.1109/TMI.2010.2043259. 18

[98] Zhang L, Li Q, You J, and Zhang D. A modified matched filter with double-sided thresholding for screening proliferative diabetic retinopathy. *IEEE Transactions on Information Technology in Biomedicine*, 13(4):528–534, July 2009. DOI: 10.1109/TITB.2008.2007201. 18, 23

[99] Gang L, Chutatape O, and Krishnan SM. Detection and measurement of retinal vessels in fundus images using amplitude modified second-order Gaussian filter. *IEEE Transactions on Biomedical Engineering*, 49(2):168–172, 2002. DOI: 10.1109/10.979356. 18

[100] Giani A, Grisan E, and Ruggeri A. Enhanced classification-based vessel tracking using vessel models and Hough transform. In *Proceedings of the 3rd European Medical and Biological Engineering Conference*, Prague, Czech Republic, November 2005. IFMBE. 18

[101] Zhang M, Wu D, and Liu JC. On the small vessel detection in high resolution retinal images. In *Proceedings of the 27th Annual International Conference of the IEEE Engineering in Medicine and Biology Society*, pages 3177–3179, Shanghai, China, September 2005. IEEE. DOI: 10.1109/IEMBS.2005.1617150. 18, 23

[102] Zana F and Klein JC. Segmentation of vessel-like patterns using mathematical morphology and curvature estimation. *IEEE Transactions on Image Processing*, 10(7):1010–1019, July 2001. DOI: 10.1109/83.931095. 18

[103] Hunter A, Lowell J, and Steel D. Tram-line filtering for retinal vessel segmentation. In *Proceedings of the 3rd European Medical and Biological Engineering Conference*, Prague, Czech Republic, November 2005. IFMBE. 18

[104] Stŏsić T and Stŏsić BD. Multifractal analysis of human retinal vessels. *IEEE Transactions on Medical Imaging*, 25(8):1101–1107, 2006. 18

[105] Kyriacos S, Nekka F, Cartilier L, and Vico P. Insights into the formation process of the retinal vasculature. *Fractals*, 5(4):615–624, 1997. DOI: 10.1007/978-1-4471-0995-2_29. 18

[106] Martinez-Perez E, Hughes AD, Stanton AV, Thom SA, Chapman N, Bharath AA, and Parker KH. Retinal vascular tree morphology: A semi-automatic quantification. *IEEE Transactions on Biomedical Engineering*, 49(8):912–917, 2002. DOI: 10.1109/TBME.2002.800789. 18

[107] Hani AFM, Ngah NF, George TM, Izhar LI, Nugroho H, and Nugroho HA. Analysis of foveal avascular zone in colour fundus images for grading of diabetic retinopathy severity. In *Engineering in Medicine and Biology Society, 32nd Annual International Conference of the IEEE*, pages 5632–5635, Buenos Aires, Argentina, August 2010. DOI: 10.1109/IEMBS.2010.5628041. 19

[108] Zhu X, Rangayyan RM, and Ells AL. Detection of the optic nerve head in fundus images of the retina using the Hough transform for circles. *Journal of Digital Imaging*, 23(3):332–341, June 2010. DOI: 10.1007/s10278-009-9189-5. 19, 47

[109] Wong DWK, Liu J, Tan NM, Yin F, Lee BH, and Wong TY. Learning based approach for the automatic detection of the optic disc in digital fundus photographs. In *Engineering in Medicine and Biology Society, 32nd Annual International Conference of the IEEE*, pages 5355–5358, Buenos Aires, Argentina, August 2010. DOI: 10.5405/jmbe.30.5.10. 19

[110] Park M, Jin JS, and Luo S. Locating the optic disc in retinal images. In *Proceedings of the International Conference on Computer Graphics, Imaging and Visualisation*, page 5, Sydney, Qld., Australia, July, 2006. IEEE. DOI: 10.1109/CGIV.2006.63.

[111] Barrett SF, Naess E, and Molvik T. Employing the Hough transform to locate the optic disk. *Biomedical Sciences Instrumentation*, 37:81–86, 2001.

[112] ter Haar F. Automatic localization of the optic disc in digital colour images of the human retina. Master's thesis, Utrecht University, Utrecht, the Netherlands, 2005. 19

[113] Osareh A. *Automated identification of diabetic retinal exudates and the optic disc*. PhD thesis, University of Bristol, 2004. 19

[114] Youssif AAHAR, Ghalwash AZ, and Ghoneim AASAR. Optic disc detection from normalized digital fundus images by means of a vessels' direction matched filter. *IEEE Transactions on Medical Imaging*, 27(1):11–18, 2008. DOI: 10.1109/TMI.2007.900326. 19

[115] Ying H, Zhang M, and Liu JC. Fractal-based automatic localization and segmentation of optic disc in retinal images. In *Engineering in Medicine and Biology Society, 29th Annual International Conference of the IEEE*, pages 4139–4141, Lyon, France, August 23-26, 2007. IEEE. DOI: 10.1109/IEMBS.2007.4353247. 19

[116] Rangayyan RM, Zhu X, Ayres FJ, and Ells AL. Detection of the optic nerve head in fundus images of the retina with Gabor filters and phase portrait analysis. *Journal of Digital Imaging*, 23(4):438–453, August 2010. DOI: 10.1007/s10278-009-9261-1. 19, 59, 99, 100

[117] Sinthanayothin C, Boyce JF, Cook HL, and Williamson TH. Automated localisation of the optic disc, fovea, and retinal blood vessels from digital colour fundus images. *British Journal of Ophthalmology*, 83(4):902–910, August 1999. DOI: 10.1136/bjo.83.8.902. 19

[118] Oloumi F, Rangayyan RM, and Ells AL. Parabolic modeling of the major temporal arcade in retinal fundus images. *IEEE Transactions on Instrumentation and Measurement (TIM)*, 61(7):1825–1838, July 2012. DOI: 10.1109/TIM.2012.2192339. 20, 50, 93

[119] Ells A, Holmes JM, Astle WF, Williams G, Leske DA, Fielden M, Uphill B, Jennett P, and Hebert M. Telemedicine approach to screening for severe retinopathy of prematurity: A pilot study. *American Academy of Ophthalmology*, 110(11):2113–2117, 2003. DOI: 10.1016/S0161-6420(03)00831-5. 20

[120] Narasimha-Iyer H, Can A, Roysam B, Stewart V, Tanenbaum HL, Majerovics A, and Singh H. Robust detection and classification of longitudinal changes in color retinal fundus images for monitoring diabetic retinopathy. *IEEE Transactions on Biomedical Engineering*, 53:1084–1098, 2006. DOI: 10.1109/TBME.2005.863971. 20

[121] Matheron G. *Random Sets and Integral Geometry*. Wiley, New York, 1974. 23

[122] Serra JP. *Image Analysis and Mathematical Morphology*. Academic Press, London, 1982. 23, 27

[123] Soille P. *Morphological Image Analysis: Principles and Applications*. Springer, Berlin, Germany, 1999. DOI: 10.1007/978-3-662-03939-7. 24, 25, 27, 29

[124] Goutsias J and Batman S. Morphological methods for biomedical image analysis. In Sonka M and Fitzpatrick JM, editors, *Handbook of Medical Imaging: Medical Image Processing and Analysis*, volume 2, pages 175–265. SPIE, Bellingham, WA, 2000. DOI: 10.1117/3.831079. 24

[125] Arcelli C and Sanniti di Baja G. Skeletons of planar patterns. In Kong TY and Rosenfeld A, editors, *Topological Algorithms for Digital Image Processing*, volume 19 of *Machine Intelligence and Pattern Recognition*, pages 99–143. North-Holland, Amsterdam, The Netherlands, 1996. 29, 32

[126] Lam L, Lee S-W, and Suen CY. Thinning methodologies - a comprehensive survey. *IEEE Transactions on Pattern Analysis and Machine Intelligence*, 14(9):869–885, September 1992. DOI: 10.1109/34.161346. 29, 32

[127] Kong TY and Rosenfeld A. Digital topology: Introduction and survey. *Computer Vision, Graphics, and Image Processing*, 48:357–393, 1989. DOI: 10.1016/0734-189X(89)90147-3. 29

[128] Rosenfeld A. Connectivity in digital pictures. *Journal of the ACM*, 17:146–160, January 1970. DOI: 10.1145/321556.321570. 31

[129] Rutovitz D. Pattern recognition. *Journal of the Royal Statistical Society*, 129(4):504–530, 1966. DOI: 10.2307/2982255. 31

[130] Hilditch CJ. Linear skeletons from square cupboards. In Meltzer B and Michie D, editors, *Machine Intelligence IV*, pages 403–420, Edinburgh, Scotland, 1969. 31

[131] Yokoi S, Toriwaki JI, and Fukumura T. An analysis of topological properties of digitized binary pictures using local features. *Computer Graphics Image Processing*, 4:63–73, 1975. DOI: 10.1016/0146-664X(75)90022-2. 31

[132] Rangayyan RM. *Biomedical Image Analysis*. CRC, Boca Raton, FL, 2005. 32, 47, 50

[133] Arcelli C and Sanniti di Baja G. On the sequential approach to medial line transformation. *IEEE Transactions on Systems, Man, and Cybernetics*, 8(2):139–144, 1978. 32

[134] The MathWorks, http://www.mathworks.com/, accessed on March 24, 2009. 32

[135] Acton ST. A pyramidal algorithm for area morphology. In *Proceedings of IEEE International Conference on Image Processing*, pages 10–13, Vancouver, BC, Canada, 2000. DOI: 10.1109/ICIP.2000.899875. 34

[136] Osareh A, Mirmehd M, Thomas B, and Markham R. Comparison of colour spaces for optic disc localisation in retinal images. In *Proceedings 16th International Conference on Pattern Recognition*, pages 743–746, Quebec City, Quebec, Canada, 2002. DOI: 10.1109/ICPR.2002.1044865. 35

[137] Al-Rawi M, Qutaishat M, and Arrar M. An improved matched filter for blood vessel detection of digital retinal images. *Computers in Biology and Medicine*, 37(2):262–267, February 2007. DOI: 10.1016/j.compbiomed.2006.03.003. 35

[138] Chui CK. *An Introduction to Wavelets, Volume 1 of Wavelet Analysis and Its Applications*, volume 1. Academic Press, San Diego, CA, 1992. 36

[139] Gabor D. Theory of communication. *Journal of the Institute of Electrical Engineers*, 93:429–457, 1946. 36

[140] Manjunath BS and Ma WY. Texture features for browsing and retrieval of image data. *IEEE Transactions on Pattern Analysis and Machine Intelligence*, 18(8):837–842, 1996. DOI: 10.1109/34.531803. 36

[141] Jones P and Palmer LA. An evaluation of the two-dimensional Gabor filter model of simple receptive fields in cat striate cortex. *Journal of Neurophysiology*, 58(6):1233–1258, 1987. 36

[142] Daugman JG. Complete discrete 2–D Gabor transforms by neural networks for image analysis and compression. *IEEE Transactions on Acoustics, Speech, and Signal Processing*, 36(7):1169–1179, 1988. DOI: 10.1109/29.1644.

[143] Daugman JG. Uncertainty relation for resolution in space, spatial frequency, and orientation optimized by two-dimensional visual cortical filters. *Journal of the Optical Society of America*, 2(7):1160–1169, 1985. DOI: 10.1364/JOSAA.2.001160. 36

[144] Ayres FJ and Rangayyan RM. Design and performance analysis of oriented feature detectors. *Journal of Electronic Imaging*, 16(2):023007:1–12, 2007. DOI: 10.1117/1.2728751. 36, 37, 67

[145] Ayres FJ and Rangayyan RM. Performance analysis of oriented feature detectors. In *Proceedings of SIBGRAPI 2005: XVIII Brazilian Symposium on Computer Graphics and Image Processing*, pages 147–154, Natal, Brazil, October 2005. IEEE Computer Society Press. DOI: 10.1109/SIBGRAPI.2005.38. 36, 67

[146] Rangayyan RM, Oloumi Faraz, Oloumi Foad, Eshghzadeh-Zanjani P, and Ayres FJ. Detection of blood vessels in the retina using Gabor filters. In *Proceedings of the 20th Canadian Conference on Electrical and Computer Engineering (CCECE 2007)*, pages 717–720, Vancouver, BC, Canada, 22-26 April 2007. IEEE. DOI: 10.1109/CCECE.2007.184. 36, 67, 72, 89

[147] Oloumi Faraz, Rangayyan RM, Oloumi Foad, Eshghzadeh-Zanjani P, and Ayres FJ. Detection of blood vessels in fundus images of the retina using Gabor wavelets. In *Engineering in Medicine and Biology Society, 29th Annual International Conference of the IEEE*, pages 6451–6454, Lyon, France, 2007. DOI: 10.1109/IEMBS.2007.4353836. 67, 72, 89

[148] Rangayyan RM, Ayres FJ, Oloumi Faraz, Oloumi Foad, and Eshghzadeh-Zanjani P. Detection of retinal blood vessels using Gabor filters. In Acharya R, Ng EYK, and Suri JS, editors, *Image Modeling of the Human Eye*, pages 215–227. Artech House, Norwood, MA, 2008. 36, 67, 89

[149] Rangayyan RM and Ayres FJ. Gabor filters and phase portraits for the detection of architectural distortion in mammograms. *Medical and Biological Engineering and Computing*, 44(10):883–894, October 2006. DOI: 10.1007/s11517-006-0109-2. 36

[150] Wolfram MathWorld: Rotation Matrix. http://mathworld.wolfram.com/RotationMatrix.html. 36

[151] Rao AR and Schunck BG. Computing oriented texture fields. *CVGIP: Graphical Models and Image Processing*, 53(2):157–185, 1991. DOI: 10.1016/1049-9652(91)90059-S. 37, 40

[152] Otsu N. A threshold selection method from gray-level histograms. *Automatica*, 11:285–296, 1975. DOI: 10.1109/TSMC.1979.4310076. 41, 94

[153] Illingworth J and Kittler J. A survey of the Hough transform. *Computer Vision, Graphics, and Image Processing*, 44:87–116, 1988. DOI: 10.1016/S0734-189X(88)80033-1. 44, 47, 50

[154] Princen J, Illingworth J, and Kittler J. A formal definition of the Hough transform: Properties and relationships. *Journal of Mathematical Imaging and Vision*, 1:153–168, 1992. DOI: 10.1007/BF00122210. 44, 47, 48

[155] Hough PVC. Method and means for recognizing complex patterns. US Patent 3, 069, 654, December 18, 1962. 44, 48

[156] Jafri MZM and Deravi F. Efficient algorithm for the detection of parabolic curves. *In Proceedings SPIE Vision Geometry III*, 2356(1):53–62, 1995. DOI: 10.1117/12.198623. 44, 50

[157] Wechsler H and Sklansky J. Finding the rib cage in chest radiographs. *Pattern Recognition*, 9:21–30, January 1977. DOI: 10.1016/0031-3203(77)90027-9. 47, 50

[158] Lu W. *Hough Transforms for Shape Identification and Applications in Medical Image Processing*. PhD thesis, University of Missouri, Columbia, MO, 2003. 47

[159] Maalmi K, El Ouaazizi A, Benslimane R, Lew Yan Voon LFC, Diou A, and Gorria P. Detecting parabolas in ultrasound B-scan images with genetic-based inverse voting Hough transform. In *Acoustics, Speech, and Signal Processing, 2010 IEEE International Conference on*, volume 4, pages IV–3337–3340, May 2002. DOI: 10.1109/ICASSP.2002.5745368.

[160] Park KS, Yi WJ, and Paick JS. Segmentation of sperms using the strategic Hough transform. *Annals of Biomedical Engineering*, 25:294–302, 1997. DOI: 10.1007/BF02648044.

[161] Zhu X, Rangayyan RM, and Ells AL. *Digital Image Processing for Ophthalmology: Detection of the Optic Nerve Head*. Morgan & Claypool, 2011. DOI: 10.2200/S00335ED1V01Y201102BME040.

[162] Banik S, Rangayyan RM, and Boag GS. *Landmarking and Segmentation of 3D CT Images*. Morgan & Claypool, 2009. DOI: 10.2200/S00185ED1V01Y200903BME030. 47

[163] Sklansky J. On the Hough technique for curve detection. *IEEE Transactions on Computers*, C-27(10):923–926, 1978. DOI: 10.1109/TC.1978.1674971. 47, 50

[164] Rangayyan RM and Krishnan S. Feature identification in the time-frequency plane by using the Hough-Radon transform. *Pattern Recognition*, 34:1147–1158, 2001. DOI: 10.1016/S0031-3203(00)00073-X. 47

[165] Rangayyan RM and Rolston WA. Directional image analysis with the Hough and Radon transforms. *Journal of the Indian Institute of Science*, 78:3–16, 1998. 47

[166] Duda RO and Hart PE. Use of the Hough transformation to detect lines and curves in pictures. *Communications of the ACM*, 15(1):11–15, January 1972. DOI: 10.1145/361237.361242. 48

[167] Duda RO, Hart PE, and Stork DG. *Pattern Classification*. Wiley, New York, NY, 2nd edition, 2001. 50, 52, 54, 88

[168] DRIVE: Digital Retinal Images for Vessel Extraction. www.isi.uu.nl/Research/ Databases/DRIVE/, accessed December 2013. 57, 58, 92, 99

[169] Structured Analysis of the Retina. http://www.parl.clemson.edu/~ahoover/stare/ index.html, accessed December 2013. 57, 58, 127

[170] Goldbaum MH, Katz NP, Chaudhuri, and Nelson M. Image understanding for automated retinal diagnosis. In *Proceedings of Annual Symposium on Computer Applications in Medical Care*, pages 756–760, November 8, 1989. DOI: 10.1109/ICIP.1996.560760. 58

[171] Hildebrand PL, Ells AL, and Ingram AD. The impact of telemedicine integration on resource use in the evaluation ROP ... analysis of the telemedicine for ROP in Calgary (TROPIC) database. *Investigative Ophthalmology and Visual Sciences*, 50:E–Abstract 3151, 2009. 58, 127

[172] De Silva DJ, Cocker KD, Lau G, Clay ST, Fielder AR, and Moseley MJ. Optic disk size and optic disk-to-fovea distance in preterm and full-term infants. *Investigative Ophthalmology and Visual Science*, 47(11):4683–4686, 2006. DOI: 10.1167/iovs.06-0152. 58

[173] DIARETDB1 - standard diabetic retinopathy database calibration level 1. http://www2. it.lut.fi/project/imageret/diaretdb1, accessed December 2013. 59

[174] MESSIDOR: Methods to evaluate segmentation and indexing techniques in the field of retinal ophthalmology. http://messidor.crihan.fr/index-en.php, accessed December 2013. 59

[175] Retinopathy Online Challenge. http://webeye.ophth.uiowa.edu/ROC/var.1/www/ index.php. 59

[176] Hamilton Eye Institute Macular Edema Dataset. http://vibot.u-bourgogne.fr/ luca/heimed.php, accessed December 2013. 59

[177] Image Processing and Analysis in Java. http://rsbweb.nih.gov/ij/, accessed on September 3, 2008. 59

[178] Metz CE. Basic principles of ROC analysis. *Seminars in Nuclear Medicine*, VIII(4):283–298, 1978. DOI: 10.1016/S0001-2998(78)80014-2. 61

[179] Wolfram MathWorld: Distance. http://mathworld.wolfram.com/Distance.html. 63

[180] Xu J, Chutatape O, and Chew P. Automated optic disk boundary detection by modified active contour model. *IEEE Transactions on Biomedical Engineering*, 54(3):473–482, 2007. DOI: 10.1109/TBME.2006.888831. 63

[181] Rogers CA. *Hausdorff Measures*. Cambridge University Press, Cambridge, UK, 1970. 64

[182] Wolfram MathWorld: Covariance. http://mathworld.wolfram.com/Covariance.html. 65

[183] Cheng HD, Jiang XH, Sun Y, and Wang J. Color image segmentation: Advances and prospects. *Pattern Recognition*, 34(12):2259–2281, 2001. DOI: 10.1016/S0031-3203(00)00149-7. 67, 68

[184] Rangayyan RM, Acha B, and Serrano C. *Color Image Processing with Biomedical Applications*. SPIE Press, Bellingham, WA, 2011. DOI: 10.1117/3.887920. 68

[185] Oloumi F and Rangayyan RM. Detection of the temporal arcade in fundus images of the retina using the Hough transform. In *Engineering in Medicine and Biology Society (EMBS), 31st Annual International Conference of the IEEE*, pages 3585–3588, Minneapolis, MN, September 2009. DOI: 10.1109/IEMBS.2009.5335389. 93, 94, 95, 97

[186] Oloumi F, Rangayyan RM, and Ells AL. Parametric representation of the retinal temporal arcade. In *Information Technology and Applications in Biomedicine (ITAB), 10th IEEE International Conference on*, pages paper no. 64 (4 pages) in CD–ROM, Corfu, Greece, November 2010. DOI: 10.1109/ITAB.2010.5687722. 93

[187] Oloumi F, Rangayyan RM, and Ells AL. Dual-parabolic modeling of the superior and the inferior temporal arcades in fundus images of the retina. In *Medical Measurements and Applications (MeMeA), IEEE International Symposium on*, pages xxxix–xliv, Bari, Italy, June 2011. DOI: 10.1109/MeMeA.2011.5966784. 93

[188] DSP++. http://www.analog.com/en/processors-dsp/blackfin/vdsp-pp-sbf/products/product.html. 124

[189] Chiang MF, Thyparampil PJ, and Rabinowitz D. Interexpert agreement in the identification of macular location in infants at risk for retinopathy of prematurity. *Archives of Ophthalmology*, 128(9):1153–1159, September 2010. DOI: 10.1001/archophthalmol.2010.199. 125

[190] Oloumi F, Rangayyan RM, and Ells AL. A graphical user interface for measurement of temporal arcade angles in fundus images of the retina. In *Canadian Conference on Electrical*

and Computer Engineering (CCECE), Proc. IEEE Canada 25th Annual, pages 4 on CD–ROM, Montreal, QC, Canada, April 2012. DOI: 10.1109/CCECE.2012.6334929. 127

Authors' Biographies

Faraz Oloumi received his B.Sc. and M.Sc. in Electrical and Computer Engineering in 2009 and 2011, respectively, from the University of Calgary, Calgary, Alberta, Canada. He is currently a Ph.D. candidate at the University of Calgary, conducting his research on image processing techniques to extract diagnostic information in fundus images of the retina. His current interests are biomedical image processing, computer-aided diagnosis, artificial intelligence, pattern analysis, and graphical user-interface design.

Rangaraj M. Rangayyan is a Professor with the Department of Electrical and Computer Engineering, and an Adjunct Professor of Surgery and Radiology, at the University of Calgary, Calgary, Alberta, Canada. He received his Bachelor of Engineering degree in Electronics and Communication in 1976 from the University of Mysore at the People's Education Society College of Engineering, Mandya, Karnataka, India, and his Ph.D. degree in Electrical Engineering from the Indian Institute of Science, Bangalore, Karnataka, India, in 1980. His research interests are in the areas of digital signal and image processing, biomedical signal analysis, biomedical image analysis, and computer-aided diagnosis. He has published more than 150 papers in journals and 250 papers in proceedings of conferences. His research productivity was recognized with the 1997 and 2001 Research Excellence Awards of the Department of Electrical and Computer Engineering, the 1997 Research Award of the Faculty of Engineering, and by appointment as "University Professor" (2003–2013), at the University of Calgary. He is the author of two textbooks: *Biomedical Signal Analysis* (IEEE/Wiley, 2002) and *Biomedical Image Analysis* (CRC, 2005). He has coauthored and coedited several other books, including *Color Image Processing with Biomedical Applications* (SPIE, 2011). He was recognized by IEEE Canada with the award of the Outstanding Engineer medal in 2013 and the IEEE with the award of the Third Millennium Medal in 2000, and was elected as a Fellow of the IEEE in 2001, Fellow of the Engineering Institute of Canada in 2002, Fellow of the American Institute for Medical and Biological Engineering in 2003, Fellow of SPIE: the International Society for Optical Engineering in 2003, Fellow of the Society for Imaging Informatics in Medicine in 2007, Fellow of the Canadian Medical and Biological Engineering Society in 2007, and Fellow of the Canadian Academy of Engineering in 2009. He has been awarded the Killam Resident Fellowship thrice (1998, 2002, and 2007) in support of his book-writing projects.

Anna L. Ells is an ophthalmologist, with dual fellowships in Pediatric Ophthalmology and Medical Retina. She has a combined academic hospital-based practice and private practice. Dr. Ells' research focuses on retinopathy of prematurity (ROP), global prevention of blindness in children, and telemedicine approaches to ROP. Dr. Ells has international expertise and has published extensively in peer-reviewed journals.

Printed in the United States
by Baker & Taylor Publisher Services